UNDISCOVERED OIL AND GAS RESOURCES:

An Evaluation of the Department of the Interior's 1989 Assessment Procedures

Committee on Undiscovered Oil and Gas Resources
Board on Earth Sciences and Resources
Commission on Geosciences, Environment, and Resources
National Research Council

National Academy Press
Washington, D.C. 1991

NOTICE: The project that is the subject of this report was approved by the Governing Board of the National Research Council, whose members are drawn from the councils of the National Academy of Sciences, the National Academy of Engineering, and the Institute of Medicine. The members of the committee responsible for the report were chosen for their special competences and with regard for appropriate balance.

This report has been reviewed by a group other than the authors according to procedures approved by a Report Review Committee consisting of members of the National Academy of Sciences, the National Academy of Engineering, and the Institute of Medicine.

The National Academy of Sciences is a private, nonprofit, self-perpetuating society of distinguished scholars engaged in scientific and engineering research, dedicated to the furtherance of science and technology and to their use for the general welfare. Upon the authority of the charter granted to it by the Congress in 1863, the Academy has a mandate that requires it to advise the federal government on scientific and technical matters. Dr. Frank Press is president of the National Academy of Sciences.

The National Academy of Engineering was established in 1964, under the charter of the National Academy of Sciences, as a parallel organization of outstanding engineers. It is autonomous in its administration and in the selection of its members, sharing with the National Academy of Sciences the responsibility for advising the federal government. The National Academy of Engineering also sponsors engineering programs aimed at meeting national needs, encourages education and research, and recognizes the superior achievements of engineers. Dr. Robert M. White is president of the National Academy of Engineering.

The Institute of Medicine was established in 1970 by the National Academy of Sciences to secure the services of eminent members of appropriate professions in the examination of policy matters pertaining to the health of the public. The Institute acts under the responsibility given to the National Academy of Sciences by its congressional charter to be an adviser to the federal government and, upon its own initiative, to identify issues of medical care, research, and education. Dr. Samuel O. Thier is president of the Institute of Medicine.

The National Research Council was organized by the National Academy of Sciences in 1916 to associate the broad community of science and technology with the Academy's purposes of furthering knowledge and of advising the federal government. Functioning in accordance with general policies determined by the Academy, the Council has become the principal operating agency of both the National Academy of Sciences and the National Academy of Engineering in providing services to the government, the public, and the scientific and engineering communities. The Council is administered jointly by both Academies and the Institute of Medicine. Dr. Frank Press and Dr. Robert M. White are chairman and vice chairman, respectively, of the National Research Council.

Support for this study by the Committee on Undiscovered Oil and Gas Resources was provided by the U.S. Geological Survey and the Minerals Management Service, Department of the Interior under agreements 14-08-0001-A0636.

Library of Congress Catalog Card No. 91-61496
International Standard Book Number 0-309-04533-9
S364

Additional copies of this report are available from the National Academy Press, 2101 Constitution Avenue, Washington, D.C. 20418

Printed in the United States of America

COMMITTEE ON UNDISCOVERED OIL AND GAS RESOURCES

CHARLES J. MANKIN, Oklahoma Geological Survey, *Chairman*
JOHN J. AMORUSO, Consultant, Houston, Texas
GEOFFREY S. BAYLISS, GeoChem-Laboratories Inc.
DORIS M. CURTIS, Curtis and Echols, Consultants
RICHARD A. DAVIS, University of Southern Florida
LLOYD E. ELKINS, Petroleum Consultant, Tulsa, Oklahoma
ROBERT J. FINLEY, University of Texas at Austin
JOHN D. HAUN, Barlow and Haun, Inc.
HARRISON C. JAMISON, Consultant, Sunriver, Oregon
GORDON KAUFMAN, Massachusetts Institute of Technology
RICHARD D. NEHRING, NRG Associates
STEVEN E. PLOTKIN, United States Congress, Energy and Materials Program
MICHAEL PRATS, Michael Prats and Associates, Inc.
RICHARD M. PROCTOR, Institute of Sedimentary and Petroleum Geology
JOSEPH P. RIVA, Library of Congress, Congressional Research Service
ENDERS A. ROBINSON, University of Tulsa
PAUL SWITZER, Stanford University

Liaison Members
CHARLES G. GROAT, American Geological Institute
HARRY C. KENT, Colorado School of Mines

National Research Council Staff
INA B. ALTERMAN, Senior Program Officer
JACQUELINE A. MACDONALD, Editorial Consultant
HUONG DANG, Project Secretary
BARBARA WRIGHT, Administrative Secretary
JUDITH ESTEP, Administrative Secretary

COMMISSION ON GEOSCIENCES, ENVIRONMENT, AND RESOURCES

CONTENTS

The members of the Committee on Undiscovered Oil and Gas Resources wish to dedicate this volume to the memory of their colleague, Harry C. Kent (1930-1991) of the Colorado School of Mines.

EXECUTIVE SUMMARY

Periodically, geologists at the U.S. Department of the Interior (DOI) convene to assess how much undiscovered crude oil and natural gas remains in the United States. Recent events in the Persian Gulf have underscored the need for such assessments. These events have once again forced policymakers to consider seriously how domestic production can be boosted to decrease reliance on petroleum from the unstable Middle East. The DOI's resource assessments provide policymakers with a key tool for evaluating ways to increase domestic production: a projection—based on scientific methods—of the quantity of oil and gas that could be developed from domestic sources.

Methods for assessing undiscovered petroleum are relative newcomers to the field of science. The U.S. government, for example, has been compiling petroleum resource estimates only since the early 1970s (though petroleum companies compiled their own estimates before that). Consequently, petroleum assessment methods are still evolving rapidly. Petroleum assessments currently in use employ a complex interplay of geology, statistics, and economics. New research in these disciplines continually offers ways to improve the reliability of the assessments. An assessment method judged reliable for an appraisal one year may be outdated by the time another appraisal is conducted a few years later.

Because petroleum resource assessments are so important to national energy policy and because assessment methods are advancing so rapidly, it is important that the DOI's assessments be evaluated periodically. This report presents a review, conducted by the National Research Council's Committee on Undiscovered Oil and Gas Resources, of the DOI's most recent resource assessment, which was published in 1989. The following summary encapsulates the committee's review; each section of the summary corresponds to a chapter in the report.

QUESTIONS ABOUT THE DOI'S 1989 RESOURCE ASSESSMENT

When the DOI released its resource estimates, some members of the natural gas industry questioned the results. Although the DOI's estimates of undiscovered oil and gas volumes were in the same range as estimates prepared by some industry analysts, they were substantially lower than other industry analysts had expected. These analysts pointed out that the 1989 estimates were lower than estimates from the two prior DOI resource assessments, published in 1975 and 1981. The volume of petroleum discovered in the years between the assessments was not enough to explain the decline in the estimated resource volumes.

In part to respond to the natural gas industry's questions, the Secretary of the Interior asked the Committee on Undiscovered Oil and Gas Resources to "review the assumptions and procedures employed in the [assessment]." Determining the degree to which the estimates are too low or too high, or whether the numbers are correct, was beyond the committee's charge, because such a determination would require almost as much effort as the assessment itself. Instead, the committee focused on evaluating the procedures used to produce the estimates.

Much of the committee's review centered on the way in which the DOI implemented a resource assessment procedure called "play analysis." The 1989 assessment was the first comprehensive national assessment in which the DOI applied play analysis, which requires a much higher level of geologic information than methods used in previous assessments. In play analysis, petroleum fields are grouped into subsets called "plays"—families of petroleum reservoirs that have geologically similar opportunities for petroleum accumulation. By combining geological judgment and statistical analysis of the history of discovered fields in a play, analysts can determine the likelihood that the play contains additional, undiscovered fields. Thus, play analysis provides a framework within which predictions about undiscovered resources are closely tied to knowledge about discovered resources.

LIMITATIONS OF RESOURCE ASSESSMENTS

It is essential to recognize that estimates of undiscovered oil and gas resources are just that: *estimates*. They are an attempt to measure uncertain quantities that cannot be accurately known until the resource has been essentially depleted. Resource estimates should be viewed as assessed at a point in time based on whatever data, information, and methodology were available at that time. They are subject to continuing revision as undiscovered resources are converted to reserves and as improvements in data and assessment methods occur.

Some of the uncertainty associated with resource estimates is inherent in play analysis. Though methods based on play analysis are widely regarded as the best way to assess petroleum resources, play analysis necessarily incorporates the subjective judgments of the resource analysts. There are no hard and fast rules, for example, for deciding which petroleum reservoirs are similar enough to include in a particular play. Furthermore, for each play, assessors must make an "educated guess" about the probabilities of occurrence of various geologic factors necessary for petroleum accumulation (a source rock, a reservoir rock, a trap, and so forth).

In addition to uncertainties inherent in assessment methods, another source of uncertainty in resource assessments is limitations in data bases. Assessors rely on two types of data bases: one containing geologic information, the other containing production and drilling information. The detailed knowledge of regional geology required for grouping reservoirs into plays comes from geologic data bases. Knowledge about the discovery history of known oil reservoirs—necessary for the statistical analyses that project the number and size of undiscovered reservoirs—comes from production and drilling data bases. Resource assessments can be only as good as the basic data that support them, so it is essential that the data bases be as complete and accurate as possible. Unfortunately, this is not always the case; for some areas, the necessary data do not exist.

Changing technology and economics are another cause of uncertainty in resource assessments. Resource estimates never include every potential source of petroleum in their final projections. Typically, they contain technical and/or economic overlays that screen out petroleum that either is not extractable with current technology or will cost more to produce than it is worth. The 1989 DOI resource assessment, for example, includes two sets of resource estimates: one that covers all resources for which recovery technology is available (termed "technically recoverable" or simply "recoverable" resources), and another that includes only the recoverable resources that will yield a profit when produced (termed "economically recoverable" resources). Yet, recovery technology is constantly improving, and petroleum that may not be extractable one year may be developed with new production methods a few years later. Likewise, changing petroleum prices and production costs may move a reservoir that is judged uneconomic in one assessment into the category of profitable resources in a future assessment.

In summary, some of the major causes of uncertainty in resource assessments are the subjective judgments that are necessarily a part of assessment methods, the data bases employed in the assessments, the continual evolution of extraction technology, and the fluctuation of production costs and petroleum prices.

AN EVALUATION OF THE DOI'S 1989 RESOURCE ASSESSMENT

After a detailed examination of the DOI's data bases, geological methods, and statistical methods, the committee judged that there may have been a systematic bias toward overly conservative estimates. Eliminating the probable sources of this bias will improve the accuracy and credibility of future assessments.

Two agencies within the DOI oversaw different portions of the assessment: the U.S. Geological Survey (USGS) assessed oil and gas onshore and in state waters, while the Minerals Management Service (MMS) assessed resources for the Outer Continental Shelf (or OCS, the part of the U.S. offshore that falls under federal jurisdiction). Though many of the committee's findings apply to both agencies, some relate specifically to the USGS or the MMS, because the agencies used somewhat different assessment procedures.

The Committee's Findings

The committee found four key characteristics of the assessment procedure that may have biased the assessment toward overly conservative resource estimates. These characteristics were improper play definition, inadequate consideration of conceptual plays, insufficient consideration of probabilistic dependencies between variables, and the unintended imposition of economic constraints on technically recoverable resource computations.

Improper play definition: The proper grouping of like reservoirs into plays is the foundation for play analysis. Mixing dissimilar reservoirs in a play may lead to biased statistical characterization of the maturity of the play. Biased statistics, in turn, may lead to a biased assessment of the potential volume of undiscovered resources in the play. The committee found that in some cases, both the USGS and the MMS created excessively large plays that contained reservoirs with diverse systems for petroleum deposition. For example, in the Pacific Coast region, the MMS divided prospective petroleum fields into plays according to stratigraphy, not by the more rigorous consideration of all the factors necessary for petroleum deposition that true play analysis requires. The USGS, in areas such as the Permian Basin, combined into plays prospective petroleum fields containing mixtures of carbonate and sandstone lithologies, suggesting that diverse depositional systems were mixed in the plays. To determine how the grouping of unlike reservoirs within a play might affect the overall resource assessment, the committee examined ten USGS Permian Basin plays in detail (see Appendix C). The analysis suggests that mixing unlike depositional systems in a play may create the false impression that the play is mature in its discovery history: i.e., that a large number of its petroleum reservoirs have already been discovered. Thus, the mixing of diverse geological systems in plays may have caused the resource estimates to be low.

Insufficient consideration of conceptual plays: Conceptual plays are plays that do not contain discovered petroleum reservoirs but that geological analysis indicates may exist. Conceptual plays may be especially important for assessing natural gas supplies, because natural gas exploration is less mature than oil exploration, so many potential natural gas producing areas lack discovered reservoirs. Though the agencies incorporated conceptual plays in some areas, the committee found that conceptual plays were not given adequate weight. For example, the MMS, in evaluating northern California's Eel River Basin, analyzed only four conceptual prospects, compared with 92 "identified" prospects. Because the information used to evaluate conceptual plays is sketchier than the data for plays with known reservoirs, conceptual plays require a separate, much more subjective evaluation procedure. Nevertheless, overlooking the importance of conceptual plays discounts a significant fraction of the nation's undiscovered resources and may have caused the estimates to be overly conservative.

Inadequate consideration of probabilistic dependence: When two unknown events are unrelated (i.e., probabilistically independent), the probability that both events will occur can be determined by multiplying the individual probabilities of occurrence of each event. However, when two unknown quantities are related (i.e., dependent), computing the probability that both will occur is more complex. In several cases, both the USGS and the MMS treated variables as independent when they may have been dependent. For example, USGS analysts assumed that the geologic events that lead to petroleum accumulation—source, timing, migration, and reservoir formation—are probabilistically independent. They multiplied the probabilities of occurrence of each of these events together to yield the marginal probability that a play contains petroleum. In most plays, however, the events necessary for petroleum accumulation are related geologically and therefore may be probabilistically dependent. No careful statistical studies of this hypothesis have been conducted, but treating such factors as independent may yield a lower probability that the play contains petroleum than if the factors were treated as dependent. Thus, the assumption that these uncertain quantities are independent when they may be dependent could have caused the estimates to be low.

Imposition of economic constraints on recoverable resource calculations: Recoverable resource estimates should be independent of economic constraints, because such estimates provide a prediction of how much petroleum could be discovered if economics were not a factor. However, the committee found that the MMS may have imposed economic constraints on its recoverable resource calculations. In Alaska, for example, MMS assessors excluded prospects that were smaller than one-half a leasing block from their technically recoverable resource estimates. This exclusion contains an implicit economic assumption: that prospects smaller than one-half a leasing block are too small to yield profits. This implicit economic assumption results in lower technically recoverable resource estimates. Also, when explicit economic screens are applied to technically recoverable resource volumes that were calculated with implicit economic assumptions, the result may be unintended double discounting and a reduction of the economically recoverable resource estimates.

In addition to finding that these four methodological limitations may have biased the estimates toward low values, the committee found that by confining the assessment to "conventional" resources, the DOI overlooked a significant portion of the potential domestic energy supply. The assessment covered only potential petroleum reservoirs that could be tapped with conventional technology. While unconventional oil deposits (like those from tar sand and oil shale) have contributed little to domestic production, unconventional natural gas deposits from low-permeability sandstones, fractured shale, and coal beds are making an important contribution to current production, thanks to new technology. Because unconventional natural gas is potentially so important to future production, furnishing estimates of unconventional gas would have painted a more realistic—and optimistic—picture of the nation's undiscovered petroleum supply.

Other Opportunities to Improve Future Assessments

In its review, the committee looked beyond factors that could have affected the magnitude of the estimates. The committee evaluated the training of staff, the integration of the diverse opinions of geologists into the final assessment results, the use of subjective judgment, the completeness of data bases, and the consideration of in-place resources.

Staff training in statistical methods could improve understanding of key probability concepts. According to a senior USGS manager whom the committee interviewed, for the 1989 assessment staff geologists' knowledge of statistics was, in many cases, limited to what they had learned in a "short course" just before the assessment. At the USGS, a permanent group of employees devoted to conducting resource assessments could help implement a statistical training program. (The MMS, unlike the USGS, maintains a permanent resource assessment staff, because it is required by Congress to perform resource assessments biennially.)

Greater consideration of the differing opinions of geologists could reflect more accurately the uncertainty inherent assessments. At both agencies, several experts' evaluations of the aggregate resources in each play were consolidated into one consensus result for the 1989 assessment. Consensus was used to obtain both the probability that each play contains petroleum and the distributions showing the range of possible resource volumes. It is inevitable that different individuals with comparable knowledge will assign different probabilities to the likelihood that petroleum is present in the same situation. Consolidating these divergent opinions into one consensus result may have artificially constricted the range of the resource estimates.

Limiting the reliance on subjective judgment could reduce some of the uncertainty in future assessments. USGS assessors used subjective judgment to extrapolate the number and size of undiscovered fields from available data about discovered fields. For analyzing plays where a moderate to large number of fields have already been discovered, there are objective models (called "discovery-process" models) available that could provide projections of undiscovered oil and gas unencumbered by judgmental uncertainties.

Expanding geologic data bases could facilitate the correct application of play analysis in some areas. Without adequate geologic data, defining plays correctly and identifying conceptual plays is difficult. The committee found that the USGS assessment was limited in particular by a lack of seismic data for the lower 48 states and also for portions of Alaska's North Slope. While the MMS's seismic data base was more complete, there were important gaps in the MMS's geochemical data. The committee found that data available from state geological surveys, state regulatory agencies, and private-sector sources (including industry) could have helped fill some of the gaps in the agencies' data bases.

Finally, expanding the assessment boundaries beyond the limits of recoverable resources could decrease assessment sensitivity to technological advances. The 1989 assessment included no estimates of in-place resources—the total volume of petroleum trapped within each play, without consideration of whether or not the petroleum is extractable. Improvements in reservoir characterization, drilling, and completion continually increase the percentage of a reservoir's total petroleum supply that is recovered. Consequently, estimates that are based on recovered resources, instead of in-place resources, become outdated as technology progresses.

RECOMMENDATIONS

Estimating how much oil and gas remains to be discovered is necessarily an inexact process. Without actually drilling, one cannot know precisely what volume of petroleum a prospective reservoir contains. Nevertheless, new assessment methods developed since the early 1970s have the ability to increase the reliability of resource assessments if properly employed. In evaluating the USGS's and MMS's assessment procedures, the committee undertook to provide the agencies with recommendations that, if implemented, will increase the level of confidence in their future assessments. The committee divided its recommendations into five categories. The first category, Assessment Boundaries, addresses how the agencies can ensure that every potential petroleum source is included in future assessments. The second category, USGS and MMS Management, suggests ways the agencies' managers can increase confidence in the assessment process. The third category, Geological Approaches in Data Base Use and Play Analysis, suggests how the agencies' geologists can upgrade their data bases and play analysis techniques. The fourth category, Statistical

Methods, recommends ways to ensure the proper application of statistics. The final category, Assessment Results, recommends ways to report future assessments so that users understand better the uncertainty inherent in the results. Below is a list of specific recommendations in each of these categories.

Assessment Boundaries

- The DOI should include estimates of natural gas from unconventional sources (separate from the estimates of conventional gas) in future resource assessments.
- The DOI should include estimates of in-place resources in future assessments. Assessors should estimate a play's in-place resource volume first, and then calculate the play's recoverable resource volume by applying a recovery factor to the in-place value.

USGS and MMS Management

- The USGS should establish a group of specialists to design and oversee on an ongoing basis a program for improving and implementing oil and gas assessment methodologies. This permanent assessment group should emphasize data validation, training of geologists in assessment methods, and more aggressive use of modern statistical methods.
- Managers at the USGS and the MMS should establish complete, continued cooperation between the assessment groups at the two agencies.

Geological Approaches in Data Base Use and Play Analysis

- Both the USGS and the MMS should conduct an audit of their drilling, geological, and geophysical data bases. The audit would have three purposes: (1) to evaluate the data's accuracy and completeness; (2) to identify areas where the data base requires improvements; and (3) to provide explicit measures of the data's quality to assessment users.
- Based on the results of the data evaluation, both the USGS and the MMS should attempt to expand and refine the available data for areas where their present data bases are incomplete. Better use of existing data bases should

precede the creation of extensive new data bases. The agencies should seek data from outside sources like state geological surveys, state regulatory agencies, other federal agencies, and the private sector. For example, it is possible that the USGS could obtain access to proprietary seismic lines from industry in key areas.

- To expand its geochemical data base, the MMS should ensure that full geochemical evaluations are conducted for wells drilled in offshore areas where existing data are inadequate.
- The USGS should analyze play content to determine the impact of play formulation—especially formulation of plays with diverse depositional systems—on the resource volumes reported in the 1989 assessment. In future assessments, the USGS should avoid creating excessively large or geologically diverse plays.
- The MMS should define plays more carefully to avoid the mixing of diverse geological and reservoir engineering characteristics in one play. The MMS should recognize that the availability of an extensive seismic data base could lead toward plays excessively dependent on recognition of structural traps. Once it has formulated plays appropriately, the MMS should institute testing to ensure that play mixing does not significantly alter resource estimates.
- Both the USGS and the MMS should incorporate more conceptual plays in future assessments. Analyzing conceptual plays may require that the agencies develop assessment techniques different from those used for known plays.

Statistical Methods

- The USGS should use objective models based on the discovery process in place of subjective extrapolations in areas with sufficient discovery data.
- To avoid unintended double discounting, the MMS should develop methods for separating technically recoverable resource calculations from those that determine the volume of economically recoverable resources.
- Both the USGS and the MMS should conduct statistical studies of risk factors, field-size distributions, prospect drilling outcomes, and other play attributes to determine whether assumptions of probabilistic independence are justifiable.
- Both the USGS and the MMS should consider ways to carry diversity of opinion through to the final resource estimates. For example, the agencies could report outlier opinions that, while masked by the aggregated results, would lead to significantly different resource estimates.

• Both the USGS and the MMS should develop explicit standards for estimating the risk that a play contains petroleum. The agencies should train their assessment staffs in the use of these standards so that different assessors use a common frame of reference when assigning risk numbers. The standards should be designed so that they can be checked against the available geologic data and replicated by other assessors.

• Both the USGS and the MMS should periodically train their oil and gas geologists in subjective probability assessment. The training should be more extensive than a "short course" just prior to the next national assessment and should be focused on real case histories.

Assessment Results

• In reporting future assessments, both the MMS and the USGS should place more emphasis on the range of uncertainty in their resource estimates. For example, the agencies could create more graphic displays to demonstrate visually the ranges of uncertainty.

• The USGS and the MMS should take special care to insure that assessment users understand the relative role that undiscovered resources play in the resource base. The agencies should explain and emphasize the undiscovered resource base's relative share as a source of reserve additions.

Incorporating the committee's recommendations in future resource assessments is likely to result in significant changes in the estimates of undiscovered oil and gas. The committee believes the overall impact will be to increase the estimated resource volumes. The DOI's procedures for the 1989 assessment appear to have been overly conservative, and therefore the assessment may have understated undiscovered resource volumes.

Implementing these recommendations requires allocation of resources over an extended time period. The lack of a permanent resource assessment group at the USGS has constrained the ability to address resource assessment methodological problems with the vigor they deserve. Unless concerted effort is made to update assessment methodologies, the next national assessment may raise the same questions as the most recent assessment.

1
INTRODUCTION: QUESTIONS ABOUT THE DEPARTMENT OF THE INTERIOR'S 1989 RESOURCE ASSESSMENT

The recent Persian Gulf crisis brought concerns about the domestic petroleum supply to the forefront of national attention. U.S. policymakers are debating how to stabilize domestic petroleum production—in addition to cutting demand—to halt the nation's growing import dependence. Once again, policymakers are considering seriously whether environmentally sensitive areas like the Arctic National Wildlife Refuge and offshore California should be opened for petroleum exploration.

Central to questions of whether a new emphasis on domestic petroleum production can slow imports and whether sensitive lands should be opened for development are predictions of *how much* undiscovered petroleum remains in the U.S. Periodic national resource assessments carried out by the U.S. Department of the Interior (DOI) provide the government with its key measure of undiscovered petroleum supplies. Every few years, the DOI has gathered geological information from the entire nation—both onshore and in the Exclusive Economic Zone, which extends 200 nautical miles out to sea—to estimate the volume of undiscovered crude oil and natural gas beneath U.S. territory.

The DOI's most recent undiscovered oil and gas resource estimate, published in 1989, caused concern among some in the petroleum industry, because it suggested that the volume of undiscovered oil and gas may be smaller than two prior government surveys, published in 1975 and 1981, had indicated. The 1989 estimate projected a mean undiscovered recoverable natural gas volume of 399 trillion cubic feet: 33 percent lower than the 1981 estimate, which had slightly exceeded the 1975 estimate (U.S. Department of the Interior, 1989). For crude oil, the drop was even more significant. The 1989 estimate projected a mean undiscovered crude oil volume of 49 billion barrels: 41 percent below the 1981 estimate, which was slightly lower than the 1975 estimate (U.S. Department of the Interior, 1989).

The apparent deviation from prior estimates caused some members of the natural gas industry to question whether the new, lower estimates were justified. An industry group, the Potential Gas Committee, had estimated undiscovered natural gas volumes close to the 1981 DOI figures (U.S. Department of the Interior, 1989). The natural gas industry was concerned that the new, lower estimates might affect government decisions about whether to provide incentives for natural gas development.

Yet, the 1989 DOI estimates may not be as low as a cursory examination of mean values suggests. The mean is only the average of a broad range of potential values, not a precise indicator of the volume of undiscovered oil or gas. The assessment concluded, with a 90 percent probability, that anywhere from 307 to 507 trillion cubic feet of natural gas and 33 to 70 billion barrels of recoverable oil remain undiscovered in the U.S. Though the mean values of these ranges are lower than the mean values from the two earlier assessments, the ranges of values for all three assessments overlap. Also, the DOI's estimate was in the same range as or *higher* than estimates prepared by some industry experts, including analysts from Sohio and Shell (see Figure 18 in U.S. Department of the Interior, 1989, for a comparison between the DOI's 1989 assessment and other assessments).

To resolve the questions about the 1989 assessment, the Secretary of the Interior called upon the National Academy of Sciences to determine whether DOI analysts had used appropriate methods to produce the new estimates. The Secretary requested that the Academy "review the assumptions and procedures employed in the [assessment]." To conduct the review, the Academy convened the Committee on Undiscovered Oil and Gas Resources, composed of 17 experts in resource assessment from academia, industry, and government. This report represents the culmination of the committee's extensive review.

WHY RESOURCE ASSESSMENTS VARY

A common problem with resource assessments is that the way they are reported often underemphasizes the uncertainty inherent in the final estimates. Users of petroleum assessments (for example, members of Congress) tend to focus on only one number, the mean value, as providing a definitive answer to the question of how much undiscovered petroleum the United States possesses. The focus on the mean value is misleading. In reality, what an assessment offers is a broad range of possible values—like the 33 to 70 billion barrel crude oil range from the DOI assessment—based on the best knowledge available at the time. No two groups of experts asked to predict the volume of undiscovered oil and gas will produce exactly the same figures.

The variation in resource assessments and their overall uncertainty have many causes. One cause of the difference in reported estimates is disagreement over which types of oil and gas should be included in an assessment. For example, the 1989 DOI assessment included only "conventional" crude oil and natural gas. The DOI defined "conventional" resources as oil or gas recoverable by currently used, widely available extraction technology. The assessment excluded oil and gas that is more difficult to extract, like gas from low-permeability sandstones and fractured shales. The PGC's natural gas assessment, on the other hand, included gas from these sources because operable extraction technology exists (though it is more expensive and is used less commonly than the technologies the DOI labelled "conventional").

A second cause of variation among assessments is different economic assumptions. Assessments commonly estimate resources that are *economically* recoverable—those that will yield a profit when developed under given economic conditions. For example, the DOI assessment assumes that industry will produce petroleum only if it yields an 8 percent return or better on the total investment (an assumption used to represent standard industry practice in deciding whether to develop a petroleum field). Calculations of whether an oil or gas source is profitable hinge on a myriad of economic assumptions, ranging from the future price of a barrel of oil to the cost of transportation to major petroleum markets. Different economic assumptions can lead to significantly different estimates of how much undiscovered oil and gas is economically recoverable.

A third cause of variation among assessments is the role played by the subjective judgments of the scientists who prepare the estimates. Subjective judgment plays a role in assigning probabilities to "risk factors" in order to calculate the likelihood that oil or gas will be present in a setting with certain geological characteristics. Accumulation of petroleum in significant quantities requires the interplay of many complex geologic events: the accumulation of organic matter in a source rock; the maturation of this organic matter into petroleum; the presence of a reservoir rock with adequate thickness, porosity, and permeability; the migration of the petroleum into a trap with adequate size and seals; and the preservation of the petroleum in the trap. Given the same set of geologic data about an oil and gas province, experts may disagree about the likelihood that each of these factors is adequate to have promoted the formation of an oil or gas accumulation.

For these and other reasons, two groups of equally trained experts can produce significantly different resource assessments. Thus, it is predictable that DOI assessments produced several years apart would yield different resource estimates, even when adjusted for petroleum discovered between the two assessments.

NEW ASSESSMENT METHODS

The 1989 assessment departed from the 1981 assessment in more ways than the final outcome. It was the first collaborative effort between two agencies within the DOI: the Minerals Management Service (MMS) and the U.S. Geological Survey (USGS). The USGS alone was responsible for the 1981 assessment. But in 1982, the DOI created the MMS to manage oil and gas resources on the Outer Continental Shelf (or OCS, the part of the U.S. offshore under federal jurisdiction). In the 1989 assessment, the USGS was responsible for estimating undiscovered petroleum volumes onshore and in state waters, while the MMS produced the estimates for the OCS. The two agencies worked fairly independently, combining their results at the end of the project.

More importantly, the 1989 assessment departed from earlier assessments in that it was the first time USGS geologists applied an assessment method called "play analysis" on a national scale. This method evaluates resource potential by grouping targets for oil and gas discovery into "plays"—families of prospective and/or discovered petroleum pools that share a common history of oil or gas generation, migration, reservoir development, and trap configuration (Podruski *et al.*, 1988; White, 1980). In play analysis, statistical methods are used to translate the judgments of geologists into a set of probabilities that given petroleum volumes will exist within the plays.

To review the assessment, the Committee on Undiscovered Oil and Gas Resources met with MMS and USGS representatives and held working sessions to evaluate aspects of the assessment. Prior to completing this review, the committee produced two related reports, reprinted in appendixes A and B. The first report evaluates the MMS's data base for assessments of resources off the California and Florida coasts. The second report examines the data base the

MMS used to estimate resources in the George's Bank area, in the North Atlantic OCS off the New England coast.

The following chapters present in detail the committee's evaluation of the USGS's and MMS's assessment methods. Chapter 2 discusses limitations of resource assessments in general. It elaborates on why uncertainty is inherent in resource assessments and why different assessments can produce such different results. Chapter 3 evaluates separately the USGS and MMS assessment methods. It offers detailed suggestions for how the assessment methods might be improved. Chapter 4 summarizes the committee's major conclusions and recommendations.

REFERENCES

U.S. Department of the Interior, Geological Survey and Minerals Management Service. 1989. Estimates of Undiscovered Conventional Oil and Gas Resources in the United States—A Part of the Nation's Energy Endowment. Washington, D.C.: U.S. Government Printing Office.

Podruski, J. S., J. E. Barclay, A. P. Hamblin, P. J. Lee, K. G. Osadetz, R. M. Procter, and G. C. Taylor. 1988. Conventional Oil Resources of Western Canada. Geological Survey of Canada Paper 87-26. Ottawa, Canada.

White, D. A. 1980. Assessing oil and gas plays in facies-cycle wedges. American Association of Petroleum Geologists, Bulletin 64: 1158-1178.

2
LIMITATIONS OF RESOURCE ASSESSMENTS

It is essential to recognize that estimates of undiscovered oil and gas resources are just that: *estimates*. They are an attempt to quantify something that cannot be accurately known until the resource has been essentially depleted. For that reason, resource estimates should be viewed as assessed at a point in time based on whatever data, information and methodology were available at that time. Resource estimates therefore are subject to continuing revision as undiscovered resources are converted to reserves and as improvements in data and assessment methods occur.

Historically, estimates of the quantities of undiscovered oil and gas resources expected to exist within a region or the nation have been prepared for a variety of purposes using several different methods. To make effective use of such estimates, or to compare them with others, one must develop an understanding of how and why they were prepared; the extent and reliability of the data upon which they are based; the expertise of the assessors; the implications and limitations of the methodology used; and the nature of any geographic, economic, technologic, or time limitations and assumptions that may apply. It is equally important that those who prepare estimates provide documentation adequate to allow the users to evaluate the issues just described. The purpose of this chapter is to examine, in general terms, some of these issues and how they may impact on the credibility and usefulness of resource estimates.

ASSESSMENT OBJECTIVES

Resource estimates serve many purposes. They may be prepared simply to inventory various energy commodities to evaluate future supply options. They provide data essential for appraising state or federal lands prior to leasing or sale. They may be undertaken to compare the relative merits of oil and gas development versus other uses for land. Large corporations and financial institutions use resource estimates for long-term planning and analysis of investment options. Industry groups, like the American Gas Association, the American Petroleum Institute, and the Potential Gas Committee (PGC), use them as guides to the future health of their industry. Increasingly, governments and the public are looking for resource estimates to provide objective statements of how much oil and gas will be available for future domestic consumption.

It is important to understand the purpose underlying any estimate, because it controls to some extent several factors that influence the assessment's outcome: the methods used, the geographic scope, the economic assumptions, the skills and biases of analysts, and the conclusions about the timing of resource availability. For some purposes, methods such as the simple extrapolation of historic resource discovery rates without regard for economic or technological change may suffice. Other uses of resource estimates may require methods capable of estimating the size ranges of individual accumulations and their reservoir properties to analyze future supply economics. Certainly if the objective of a national resource estimate is to determine if sufficient oil and gas will be available to maintain an acceptable standard of living and a viable industrial complex, then one must be concerned about levels of accuracy and the adequacy of data and analysis. One must be assured that all potential oil and gas sources have been considered. Reactions of citizens and governmental bodies, and resulting public policy, will be quite different if there is a perception of future domestic abundance rather than a perception of future domestic shortage; consequently, the need for accuracy is paramount in national resource estimates designed to help determine public policy.

UNCERTAINTY IN ASSESSMENTS

Uncertainty is an integral part of resource estimates. Almost every component of the assessment process has associated uncertainty, and the aggregate level of uncertainty for the final resource estimates can be large. Major uncertainties arise from limitations in the data base and methodology, difficulties in projecting the course of recovery technology, and the sensitivity of the economic analysis to changes in energy prices and production costs.

In general, uncertainties in estimates of undiscovered oil and natural gas are greatest for areas that have had little or no past exploratory effort. For areas that have been extensively explored and are in a mature development stage, many of the unknowns have been eliminated and future resources can be evaluated with much more confidence. Even in some mature producing areas, however, uncertainty remains about the potential oil and gas supply at greater drilling depths. Uncertainty also pervades projections of whether potential reservoirs have been unrecognized or bypassed in past drilling. Similarly, where resource estimates are based on analogue comparisons between maturely explored areas and unexplored areas, uncertainty is introduced because each area or basin has unique characteristics.

Although our fundamental knowledge of the origin, migration, and entrapment of oil and gas has advanced markedly during the past 30 years, the fact that incremental scientific advances are still being made leads to additional uncertainty in resource estimation, especially in frontier areas or at great drilling depths. In other words, new knowledge may lead to increases or decreases in estimates of undiscovered resources, but generally leads to a reduction of uncertainty.

Discovery is only the first step in crude oil and natural gas resource development. The present state of technical knowledge in reservoir geology and petroleum engineering, as well as existing regulations, determine the spacing, completion, and production methods of development wells (i.e., petroleum producing wells drilled after the discovery well). As engineering and geologic knowledge increase, our ability to withdraw larger increments of oil and gas from existing fields is enhanced. Thus, the sizes of fields in terms of ultimately producible barrels of oil or cubic feet of gas increase with time. Uncertainty as to ultimate sizes of *discovered* fields leads to uncertainty in estimates of size distributions of *undiscovered* fields in areas with analogous reservoir characteristics and geologic histories.

Scientists can estimate the quantity of technically recoverable undiscovered oil and gas based on the present state of geological and engineering knowledge, modified by a consideration of future technological advancement. However, the percentage of that quantity that may actually be discovered and produced is an economic question. Uncertainties about future well-head crude oil and natural gas prices and costs of exploration and development adversely affect all resource estimates. In short, uncertainties embodied in economic assumptions lead to significant uncertainties in estimates of economically recoverable resources and account for some of the large differences among estimators.

There are no foolproof, completely mechanical methods for estimating undiscovered resources. Because all methods contain elements of subjective judgement or expert opinion, the degree of uncertainty is affected by the level of expertise of the personnel doing the estimating, the time devoted to the estimates, the methods by which the estimates are tested by peer review, and the level of enthusiasm for the appraisal.

ASSESSMENT BOUNDARIES

In general, policy concerns about oil and gas resources are actually concerns about future domestic production rates. Policymakers tend to view resource estimates as indicators of production capability. However, the results of resource assessments are not often presented in a form that recognizes this concern with future production. Instead, the results commonly represent only a piece of the resource base.

The USGS/MMS assessment evaluated in this report covers *conventional undiscovered recoverable* oil and gas—only a part of the total resource supply that will contribute to future production. To the uninformed user of this assessment, it might seem that what the assessment includes is straightforward: undiscovered petroleum recoverable with existing technology. Yet, in this and other resource assessments, the distinctions between "conventional" and "unconventional," "recoverable" and "unrecoverable," "discovered" and "undiscovered" are hazy. To interpret the results of an assessment correctly, the user of the assessment must understand where the assessment's boundaries are drawn—what is and what is not included in the assessment.

Discovered/Undiscovered Boundary

To some, the difference between undiscovered and discovered petroleum might seem obvious. The term "undiscovered" implies a clear distinction between resources we have found, identified, and measured and those we have not. Yet, in some cases, deciding what to classify as discovered is a judgment call that depends on an analyst's interpretation of oil-industry definitions.

Resource appraisers divide the nation's remaining oil and gas reserves into four categories, familiar to those in the petroleum industry:

• *Measured reserves* are contained in known reservoirs, have been quantified by engineering studies, and are recoverable under existing economic and operating conditions.

• *Indicated reserves* are reserves from known reservoirs for which extraction depends upon improved recovery techniques.

• *Inferred reserves* are as-yet undocumented resources that analysts expect may be added to existing petroleum fields from extensions and additional development.

• *Undiscovered resources* are those outside of known petroleum fields that analysts postulate to exist based on broad geologic knowledge.

Figure 2.1 diagrams the distinctions between these types of reserves. On the figure, level of certainty that the reserves exist increases from right to left.

The distinction between inferred reserves and undiscovered resources may confuse users of assessments. Depending on how an assessor interprets the above definitions, a postulated but unknown reservoir that a layman might believe is "undiscovered" can be defined as "inferred" and therefore categorized as "discovered."

Conventional/Unconventional Boundary

The term "conventional" implies that a common method of resource extraction or a well-identified set of physical characteristics clearly separates "conventional" from "unconventional" petroleum. This is at best an oversimplification. The boundaries between conventional and unconventional resources have always been somewhat indistinct and have become increasingly so over the past several years. This complicates comparisons between alternative resource assessments and interpretation of individual assessments. Problems with maintaining strict boundaries include:

• Relatively large quantities of natural gas from sources traditionally labelled "unconventional" are being produced in some areas. Examples of "unconventional" resources whose production is becoming more common are coal-seam methane and gas from low-permeability sandstone reservoirs (called "tight gas") and fractured shale reservoirs.

• Technologies used to produce unconventional gas, especially fracturing, are now being widely used in formations that are considered "conventional."

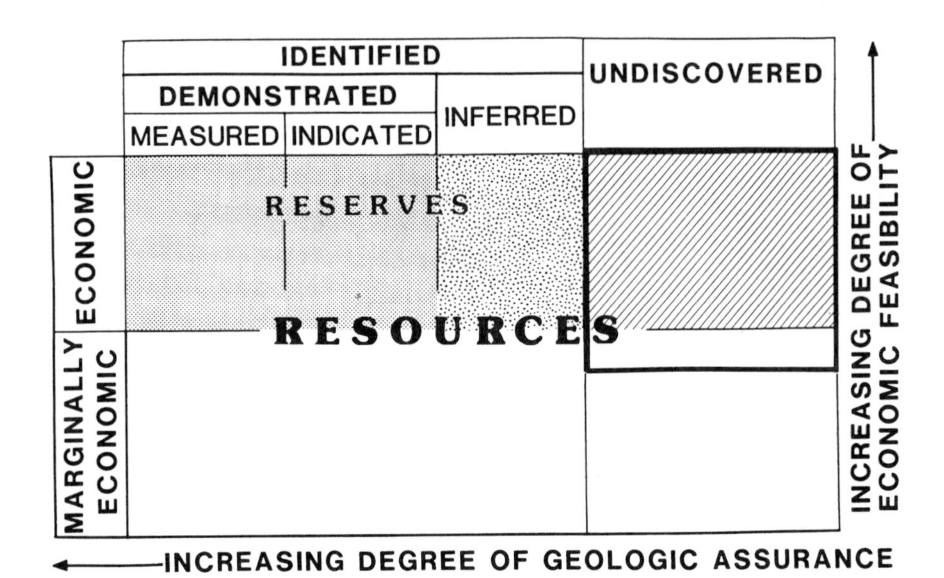

FIGURE 2.1 A diagram showing how petroleum resources are classified. The area within the heavy frame on the upper right represents the undiscovered recoverable resources estimated in the 1989 DOI assessment. The hachured area within the heavy frame indicates undiscovered resources that are estimated to be economically recoverable. Source: U.S. Department of the Interior, 1989.

- Reservoir permeability, which is the defining characteristic for tight-gas reservoirs, can vary widely within individual formations; many formations combine permeability characteristics on both sides of any defined conventional/unconventional boundary. Similarly, gas withdrawn from coal seams may contain gas from both the seam and the adjacent sedimentary strata—that is, a combination of conventional and so-called unconventional gas.

Available data sources cannot distinguish between conventional and unconventional gas in these circumstances.

In the past, analysts placed resources in the "unconventional" category for two reasons: to designate resources that were considered a potential but not a current source of production (except perhaps in small quantities), and to identify resources that required production methods significantly different from those normally used. Thus, as production of at least a portion of the unconventional resources has grown, and as the technology for this production has become widely used in the industry and thus "conventional," it makes sense to move some of the unconventional resources out of this category and into the category of conventional. Allowing the boundaries between conventional and unconventional resources to be frozen according to obsolete conventions and practices fails to recognize important technological advances.

Recoverable/Unrecoverable Boundary

Most assessments use two criteria to define the boundary between recoverable and unrecoverable resources: technology and economics. *Technically recoverable* resources are resources for which technology exists that can locate and produce the oil and gas and transport it to market. *Economically recoverable* resources are those that a profit-driven industry would seek to exploit. (The USGS/MMS assessment gives two estimates of undiscovered oil and gas reserves: one that includes resources recoverable with conventional technology and one that includes only the recoverable resources that yield an 8 percent real profit.) The problem with categorizing resources as technically or economically recoverable is that changing assumptions and changing times can drastically alter these criteria and hence the assessment results.

Technically Recoverable Resources

Resource assessments should, but rarely do, begin with a calculation of *in-place resources*: the *total* volume of petroleum trapped within the play being assessed, without consideration of whether or not the petroleum is extractable. Instead, assessments typically limit their evaluation to resources recoverable with conventional technology. Thus, when assessors consider what resource volume exists in a petroleum field, they weight their considerations with a judgment about how much of the resource is recoverable. In part, this is because until recently, obtaining data on in-place resources was difficult, as this chapter will discuss later.

The distinction between in-place resources and recoverable resources arises because a substantial part of the earth's total petroleum content is either dispersed at very low concentrations throughout the crust or is present in forms that cannot be extracted except through methods that would cost more than the petroleum is worth. An additional portion of the total petroleum resource base cannot be recovered because available production technology cannot extract all of the in-place crude oil and natural gas even when these resources are present in commercial concentrations. This is particularly important for crude oil; using current practices, only an average of one-third of the total in-place crude oil in a reservoir can be recovered by primary and secondary recovery processes. The remaining two-thirds of the resource may not be recoverable at the present time, but it does remain a target for future technology. For natural gas, a higher percentage (60 to 85 percent) of in-place reservoir natural gas is recovered using current technology.

The inclusion of oil and gas resources in the category of technically recoverable or technically unrecoverable is at best a snapshot in time, and, if history is to judge, one destined to be quickly rendered obsolete. The development of improved technology continually causes resources to move from the unrecoverable to the recoverable portion of the resource base. Improvements in offshore drilling technology have allowed drilling in deeper waters and more hostile conditions, opening up new territories to development. As a result, succeeding generations of natural gas resource assessments have steadily moved their boundaries to deeper and deeper waters. Other advances in offshore technology have lowered costs sufficiently to allow the exploitation of fields previously considered too small to develop profitably; thus, technical advances may move resources across economic as well as technical boundaries. Refinements in seismic techniques have allowed exploitation of the Western Overthrust Belt, which has very complex natural gas-bearing formations. (Here, there may be arguments about whether the boundary crossed was technical or economic, since intensive conventional drilling might have revealed the gas fields in this area.) Progress in horizontal drilling technology has opened production zones that otherwise would be uneconomic. Improvements in drill bit metallurgy, downhole drill control, and drilling mud systems have allowed exploration of deep natural gas deposits unreachable by earlier, "conventional" drilling technology.

The language describing technology assumptions in current resource assessments is sufficiently vague to inspire doubts about just where the technological boundaries lie. The Potential Gas Committee (PGC) assumes the use of "current or foreseeable technology." The DOI defines its technological boundary by exclusion: it excludes all resources except those that can be produced by "natural pressure, pumping, or injection of water or gas." It does not mention exploration technology, even though an inability to locate resource deposits in complex formations can prevent recovery.

Our concerns about precisely where the technological boundaries of an assessment center on the following issues:

- How does the assessment treat resources that theoretically could be recovered with conventional technology used at an unusual degree of intensity (for example, drilling at a closer-than-standard spacing) but that are unlikely to be economically recovered without development of new technology?
- Many technologies that are "commercial" remain the province of a small portion of the potential users or are restricted to a few geological areas. Does the assessment assume that these technologies will be in use universally?
- Just what does "the foreseeable future" mean?
- In the absence of specific assumptions about exploration technology, does the assessment include all oil and gas fields containing economically recoverable resources—regardless of whether the technology for finding them is economical or available?

With the exception of a limited effort by the PGC, no assessors have attempted to identify the extent to which improved technology has caused resource estimates to increase over the years. Without the availability of such information, it is difficult to quantify the effects of technological improvement on total recoverable resource volumes. Consequently, it is difficult to predict credibly how future assessments are likely to change with technological improvements. However, we can identify the advances that are most likely to increase recoverable resource volumes in the near future. These include:

- extension of horizontal drilling to common use, which will boost accessible resource volumes in thin pay zones and in highly fractured and compartmentalized reservoirs;
- advances in chemical-enhanced oil recovery methods, which will raise recovery efficiencies in fields undergoing tertiary recovery;
- progress in reservoir characterization and modelling, which should enhance the potential for using infield drilling to add to reserve growth; and
- improvements in subsea completions, which will allow the development of smaller fields and fields farther offshore and in deeper water than was previously possible.

Economically Recoverable Resources

Of the technically recoverable resource, assessors wish to eliminate resources that exist in volumes too small, in waters too deep, in strata too far underground, in places too remote, or in physical conditions too difficult to allow for profitable extraction and shipment. Users of the assessment must recognize, however, that the location of the economic boundary will depend on both the economic assumptions chosen and economic conditions at the time of the assessment.

Economic boundaries are introduced into resource assessments in one of three ways: (1) through quantitative analysis using *explicit* assumptions about resource prices, economic conditions, and available technology; (2) through quantitative analysis using *implicit* economic assumptions like minimum economic field size (the smallest size field assessors determine is profitable to develop) and water depth limits; or (3) through subjective, qualitative assumptions about economic relationships. An example of (3) is the PGC's definition of economic boundaries: "only the natural gas resource which can be discovered and produced using current or foreseeable technology and under the condition that the price/cost ratio will be favorable." With (1) and (2), assessors attempt to define precise limits of economic attractiveness. They incorporate assumptions that may include an explicit price scenario that varies over time, a statement of the tax regime under which resource exploration and production take place, and a minimum acceptable rate of return. The MMS and USGS used this latter approach in their assessment.

Defining the economic boundaries of a resource base is a controversial issue in resource assessment, primarily because the economic attractiveness of resources is volatile over time and depends on a complex set of variables, including the state of technology. Furthermore, economic analysis undertaken during the past few years has been especially problematic for oil and gas resource assessment. The events of the past 15 years have demonstrated the instability of both costs and prices: the rapid "boom" in oil-industry activity from the late 1970s until about 1981, followed by the gradual decline from 1981 until 1985, followed by the plunge in 1985 and 1986 in both world oil prices and activity levels.

During the oil price increases of the 1970s and early 1980s, drilling and service costs rose sharply and industry efficiency fell. As a consequence, the actual price-driven expansion of the recoverable resource base was considerably below what might have occurred had costs and efficiencies remained constant. Similarly, although the gradual and then sharp oil price drops of the 1980s pushed many prospects out of the range of economic viability, the effect was smaller than expected. Large declines in drilling and field-service costs have moderated the negative impact of the price drops, limiting the declines in operator profitability. In addition, the economic pressure that oil and gas operators experienced during the past few years stimulated major efforts at cost reduction and increased efficiency through better selection of prospects. These efforts have further alleviated the effects of the price drop. Thus, it has become common to learn about prospects being developed at $18 per barrel of oil that a few years ago were believed to require $25 per barrel or more to develop profitably.

The positive relationship between oil and gas prices and drilling and service costs, described above, implies that resource-base boundaries will be less sensitive to resource prices than would be expected from economic analyses assuming constant costs. Nevertheless, changes in resource prices or in other economic conditions—such as tax rates or allowed tax credits or income deductions—can add to or subtract from the economically recoverable resource. Users of any resource assessment should be aware of the nature of the assumptions that define the resource boundaries.

In addition to price fluctuations, another factor that strongly affects the economics of oil and gas resource development, and thus the volume of oil and gas within an assessment's "economically recoverable" boundary, is the available infrastructure: pipelines, ports, gas processing plants, and so forth. Where a strong infrastructure already exists, resources that would otherwise be uneconomic—those located in small fields or in difficult and expensive operating environments—often may be developed profitably. Most areas within the onshore lower 48 states have a strong existing infrastructure.

In more remote areas, however, the decision to develop must be predicated on having to invest heavily in infrastructure. For example, in many areas of Alaska's North Slope, development of future discoveries cannot proceed without building a pipeline to the Trans Alaskan Pipeline System and constructing extensive port and airfield facilities. For such an investment to be profitable, explorers must find large fields, with oil volumes of several hundred million barrels. In addition, natural gas is excluded from the economically recoverable resource base, because no pipeline exists to transport the natural gas to markets in the lower 48 states.

Assessment users should understand that once infrastructure is constructed, future assessments can change drastically. For example, if a pipeline were built from Prudhoe Bay to the Arctic National Wildlife Refuge, future resource assessments would likely incorporate substantial new resource volumes from fields smaller than the smallest field that was necessary to support the initial pipeline development.

DATA BASES FOR RESOURCE ASSESSMENTS

Two types of data bases play a key role in preparing resource estimates. One describes the geologic framework and is the starting point for all estimates. The second includes all of the information generated as wells are drilled and discoveries are made. Because resource assessments can be only as good as the basic data that support them, it is essential that the data bases be as complete and accurate as possible. Unfortunately, this is not always the case. And, assessors commonly understate the inadequacy of their data, if they mention it at all.

Geologic Data Base

Geologic data bases typically are compiled by a group of experts in various disciplines formed to analyze the region's geologic history. Ideally, the experts pool their knowledge of the region's stratigraphy, sedimentation, tectonics, and geochemistry to develop a "best possible" understanding of area's development in terms of opportunities for petroleum accumulation. They apply knowledge from scientific literature, consult with local experts in state and federal agencies and academia, solicit advice or assistance from industry petroleum explorers, and conduct original studies to the extent that time and resources permit. Any weaknesses in the geologic data base they create will be carried through whatever methodology is used in the assessment process. Unfortunately, the quality and completeness of the geologic data base are rarely adequate. This is, in part, because assembling it is a time-consuming activity that requires the dedication of people and resources on an ongoing basis.

Compiling the geologic data base requires geophysical data, particularly seismic data, both regional and detailed in scope. Most data from both state waters and the onshore are the proprietary property of oil companies and are not generally available to the USGS. The MMS, on the other hand, does have access to such data in the federal OCS waters. This creates a basic disparity in resource estimation capabilities and illustrates a serious weakness in the onshore data base.

Production and Drilling Data Base

The data base related to exploration and production history commonly is compiled from various federal and state agencies, trade associations, and commercial enterprises. More than three million wells have been drilled in the United States, creating an almost overwhelming amount of data. Thousands of staff years have been expended in correlating, summarizing, evaluating, and statistically analyzing these data.

The single most important component of this data base is the estimates of *proved reserves*: petroleum from already-discovered reservoirs that is recoverable under existing economic and operating conditions. One might expect that quantitative information on reserves would be relatively complete. Unfortunately, this is not the case. One problem with reserves data is that because reserves tend to grow, it is difficult to know the appropriate ultimate reserve value for any field. A second problem is that although data on large- and intermediate-size fields are comprehensive in some instances, data for small fields are largely unavailable. Lack of data on small fields is an important shortcoming, because most remaining onshore undiscovered fields will be small relative to previously discovered fields. A third problem is that producers generally treat proved-reserve estimates as confidential. Producers aggregate by state or state subdivision the data they do release, withholding information on individual fields. There are some exceptions, but this is the general rule.

Virtually all modern methods of resource estimation depend on reliable reserves data. Some methods are essentially based on the sizes of discovered fields. It is therefore imperative that reserve values accurately reflect the real sizes of accumulations. Current data base development focuses largely on production data. That is, data bases express reserves as the component of the total accumulation deemed recoverable at any point in time. Thus, reserve estimates reflect the recovery technology and economics of the time. As such, they are subject to considerable change over time. The consequence is that reserve values can be poor facsimiles for the ultimate size of a field.

To circumvent this problem, assessors could use in-place values of resources and then apply appropriate recovery factors to estimate recoverable resources (Podruski *et al.*, 1988). (Recovery factors express the percentage of in-place petroleum that can be brought to the surface.) In the past, however, determining in-place values has been difficult because of the lack of a comprehensive public data base of U.S. in-place petroleum. Several public and private data bases covered one or more elements of in-place petroleum. However, each of these was incomplete, lacking either data on the total petroleum content of each reservoir or complete national coverage.

One example of a data base that included information on in-place petroleum was the American Petroleum Institute's (API) industry-wide crude oil data base. Unfortunately, the API stopped compiling this data base in 1979. The data, which reached back to 1920, covered annual statistics on discoveries, production, reserves added, estimated ultimate recovery, and original oil in place. The data base aggregated in-place petroleum volumes into production districts, states, and fields for the hundred largest fields in the U.S., but did not include figures for in-place reserves at other specific reservoirs.

In the future, determining in-place resource values may be easier, because the Department of Energy (DOE) is upgrading its Tertiary Oil Recovery Information System (TORIS), which provides data on original oil in place in individual reservoirs. Since its initiation in 1978, TORIS has provided comprehensive data on original oil in place in areas where substantial petroleum exploration has already taken place; TORIS data account for more than 72 percent of the original crude oil in place for the known fields of the nation. However, TORIS's coverage of less-explored regions has, in the past, been incomplete; it includes information on only 3,700 of the more than 100,000 U.S. reservoirs (Brashear *et al.*, 1989). Currently, the DOE is expanding TORIS to include more data about less-explored areas. The expansion of TORIS will help facilitate the use of in-place resource values in projecting undiscovered oil volumes.

ESTIMATING RESERVE GROWTH

Past production plus total estimated future production (from proved reserves) equals the so-called estimated "ultimate" amount of oil or gas that will be recovered from an existing field. However, predicting a field's true ultimate recovery requires an estimate of its future reserve "growth." Reserve growth can result from several factors: data reevaluation, extension drilling, new-pool exploration, infill drilling, enhanced-recovery techniques, better reservoir management, and changes in economic parameters. Some changes in reserve estimates may be negative rather than positive, but history has shown that, on average, reserve estimates grow with time. During the first five to seven years after explorers discover a field, estimates grow rapidly. Later, they tend to level out at some smaller increment per year. It is worth noting that recent annual growth of estimated ultimate oil recovery from pre-1920 fields is still one to two percent per year.

Much of the increased reserves in large fields is the result of infill drilling and advances in enhanced-recovery techniques made possible by the higher oil prices of the late 1970s and early 1980s and from the knowledge gained during that period. The need for up-to-date reserve-growth estimates stems from the knowledge that changing economic and technological factors profoundly change the ultimate quantity of oil and gas that will be produced from existing and yet-to-be-discovered fields.

Historic data can be used to project future growth of newer fields. (As discussed in the next chapter, the USGS uses such a method of trend extrapolation.) One source of historic data is the revised estimates of oil and gas discovered in the United States each year since 1920 that were published annually by the American Petroleum Institute, the American Gas Association, and the Canadian Petroleum Association from 1968 until 1980 (for the years 1967 to 1979). Analysts have used these data to develop statistical growth curves for estimating inferred reserves. Because this data series stopped in 1979, more recent reserve growth estimates based on the series fail to capture a decade of innovation in petroleum development geology and engineering.

Estimating reserve growth with direct geologic and engineering methods is also possible. For example, the PGC employs such a direct method to estimate additional natural gas resources in discovered fields. The PGC assesses accumulations thought to be only partly developed by using a yield factor and extended reservoir volume, multiplied by the probability that the extension actually exists. PGC assessment of all possible new pools within existing fields also involves volumetric calculations, with adjustments for possible changes in lithology and reservoir characteristics. Adjusted volumes are multiplied by adjusted yield factors, obtained by analogy with a producing area similar to the potential new pool under consideration. The amount of gas is then discounted by the probability that unknown traps exist and contain gas. Unfortunately, this direct approach to predicting future growth of newer fields requires detailed field information that commonly is proprietary and not readily available. Also, it is unlikely that this method captures all reserve growth that may result from new concepts of reservoir heterogeneity of already-developed reservoirs.

ASSESSMENT METHODOLOGY

A variety of methods for evaluating undiscovered oil and gas resources has evolved over the last 25 years. These methods range from simple models based on historical data to large and complex models based on probability theory. This section discusses the methods used most extensively, particularly by government agencies, in making regional and national assessments in North America during the 1970s and 1980s.

The Evolution of Government Assessments

In response to perceived global energy concerns of the early 1970s, the USGS embarked on a program to analyze the nation's oil and gas resources. Although major petroleum corporations had been conducting assessments for many years for internal purposes and had developed appropriate methods, by comparison the federal government was ill prepared for the task. The need to prepare estimates quickly, along with funding and staff limitations, led quite naturally to the acceptance of relatively crude methods. Early assessments began with a request for regional geologists to examine and summarize the petroleum geology of large regions. Analysts used those perceptions of regional geology along with comparisons to global analogues and volumetric yield factors (barrels per cubic mile of prospective basin-fill) derived from the literature to estimate ranges of possible oil and gas quantities. They employed a Delphi-like approach, averaging the opinions of several experts to produce the final results. The USGS published the results of these assessments in Circulars 725, 828 and 860.

In attempting to address the relatively simple objective of the early 1970s—to determine how much oil and gas might exist—USGS analysts recognized the need for improved methods. It was clear that a better means of systematically capturing, quantifying and reporting uncertainty was required; that regions had to be assessed in a more disaggregated fashion to use analogue information more effectively; that the objectives were more comprehensive than originally thought and ultimately must address questions of deposit size and exploration risk; that effective protocols for handling subjective judgment had to be developed; and that statistical methods and methods testing would be desirable. These concerns gave rise to increased research on appropriate methods. By the late 1970s, scientists had developed new methods that grouped similar oil and gas reservoirs into "plays" for assessment (see Box 2.1) and

BOX 2.1 How to Define a Play

In defining a play as "a group of geologically similar prospects having basically the same source-reservoir-trap controls of oil and gas," White stressed the importance of geologic commonality in play definition (White, 1980). Achieving geologic commonality is essential for insuring that each set of reservoirs and prospects being evaluated is as homogeneous as possible. Yet, in practice, achieving commonality is not a straightforward exercise. There is no single operational formula for play definition that fits all geologic settings.

In the great majority of cases, plays are limited to a single formation, because each formation is associated with a distinct set of reservoir characteristics. Yet in some areas, notably California and some Rocky Mountain basins, a single play may encompass several producing formations, the play being essentially defined by a structural trend. Within a single formation, depositional system can be a key parameter in play definition, because differences among types of depositional systems are associated with differences in reservoir size distributions and patterns of reservoir location. Differences in petroleum sources, in thermal maturity within a source, and in migration history as they affect petroleum type and characteristics can also be important factors in determining play boundaries.

applied the widely recognized concept that petroleum pool or field sizes appear to be lognormally distributed in nature (Kaufman, 1965). Most importantly, the new methods made extensive use of subjective probability. (Roy *et al.*, 1975; Roy, 1979; Energy Mines and Resources, Canada, 1977; White and Gehman, 1979).

Subjective Probability Methods

Central to the new subjective probability methods was the solution of some variant of the standard engineering equation for calculating the reserves in an individual pool or field, but applied to all of the prospects in a play. In this process, a frequency distribution appropriate for all prospects represents each variable in the equation (net pay, area, porosity, hydrocarbon saturation, etc.). These distributions incorporate all the observed values that result from discovery, supplemented with the assessment team's subjective judgment. The distributions are inserted in the standard engineering equation in place of single number values for the variables. Where the equation requires, analysts multiply the distributions together with an appropriate mathematical procedure, such as Monte Carlo simulation. The equation then produces a conditional pool-size distribution: a curve showing the possible sizes of petroleum pools plotted against their frequency of occurrence.

The pool-size distribution is called "conditional" (or, alternatively, "unrisked") because it assumes the condition that the play contains some minimum amount of petroleum. In other words, analysts consider the possible size attributes of petroleum deposits separately from the question of whether the play contains any petroleum. One can view the probability that a play contains petroleum as analogous to the chance of success of a wildcat exploratory well. Determining this probability requires knowledge of the geologic factors—a source rock, a reservoir rock, a trap, and so forth—necessary for petroleum accumulation. Based on geologic evidence, analysts assess a marginal probability for each such geologic factor. Then, they multiply the marginal probabilities together (a calculation that requires the sometimes questionable assumption that the factors are statistically independent) to obtain the chance that the play contains petroleum.

By combining the exploratory well success probability, the conditional pool-size distribution, and a second distribution representing the number of prospects in the play, analysts produce a probability curve showing the total petroleum volume in the play.

Subjective probability methods gained favor with assessors in part because they can be geology-based and because they provide a relatively simple means of reflecting uncertainties associated with the variables that describe pool size. A problem with such methods, however, is that the Monte Carlo combination of size variables is legitimate only if each variable is functionally independent. Empirical evidence has shown that this is not always the case. There may be correlations between area and net pay; porosity and depth; and other sets of variables. More recently, analysts have altered programs to address possible dependence between size variables.

A variation on the subjective probability approach forms the basis for part of the MMS's PRESTO (Probabilistic Resource Estimates—Offshore) model: a computer program for simulating exploratory drilling. The PRESTO program applies subjective probability to assess individual prospects which can then be summed into plays.

The subjective probability methods described above can be shown to be statistically valid (Lee and Wang, 1983a,b), but commonly serious problems arise in their implementation. One problem is that analysts typically have little or no training in assessing uncertainties or in evaluating probabilistic dependencies between variables. A second problem is that an assessment team commonly reaches consensus at the expense of capturing the total spread of opinion regarding uncertainty. Assessments disguise the opinions of individual assessors by combining these opinions and publishing only a single, group probability distribution. A third problem is the absence of protocols to ensure consistency in assessment methods. Lacking such protocols, assessors may interpret differently even terms that seem relatively straightforward, like "maximum value" and "minimum value." For example, one assessor may interpret the "minimum value" of the number of prospects in a play as a value such that a small but *non-zero* probability exists that the number of prospects is less than this "minimum." Another assessor may interpret the same term to mean that there is *zero* probability that the number of prospects falls below this minimum. Assessors are generally aware of the potential problems with these subjective probability methods, but feel that there is such a large error cloud involved that many of the problems can be safely ignored.

Discovery-Process Models

To sidestep the problems with subjective probability methods, assessors developed a new generation of methods that centered on objective, probabilistic models of the petroleum discovery process. Such "discovery-process models" are built from assumptions that describe both physical features of petroleum deposits and fields and the manner in which they are discovered. Their principal premise is that discovery data are size biased: large deposits are more likely to be discovered early in the evolution of a play than are small fields (Kaufman *et al.*, 1975). Thus, exploration produces a sample of the petroleum pool population that is biased toward early discovery of larger pools. By understanding the nature of the bias and modelling it mathematically, Kaufman derived the parameters describing the North Sea petroleum pool population and successfully estimated resource values for North Sea plays. Lee and Wang built on Kaufman's work, employing somewhat different mathematical procedures to derive the population parameters and adding the capability to determine individual pool sizes (Lee and Wang, 1985).

Assessment methods based on discovery-process models have two obvious advantages. First, the input required comes from two of the most readily available and reliable types of geological data: the sizes of discovered deposits and the order of discovery. Second, the assumptions defining the model can be tested for validity. A difficulty of the process, common to all assessment procedures, is that assessors must be careful in defining exploration plays. Faulty definitions can result in mixing of plays, which can seriously distort the model's projections of undiscovered oil and gas, as is shown in Appendix C.

In the assessment this report reviews, the USGS employed a hybrid of subjective probability methods and discovery-process models. A fundamental concept from discovery-process models is that a play's field-size distribution (the graph showing the frequency at which each possible field-size occurs) shifts predictably with time. As exploration progresses, on average, the proportion of large deposits remaining undiscovered decreases rapidly relative to the proportion of smaller fields; the field-size distribution shifts toward a smaller percentage of large fields. Thus, the return on investment decreases as exploration progresses, as the large fields are exploited and the less profitable small fields remain. By applying this concept and creating different field-size distributions for different points in time, USGS analysts projected the rate of falloff of returns on exploratory efforts.

Subjective probability entered into the USGS assessment in the way analysts created field-size distributions. Instead of applying objective discovery-process models, USGS analysts used subjective judgment and a function called a "Pareto" distribution: a probability curve that analysts can tailor to fit known field-size data. To create field-size distributions, analysts fit sizes of known oil and gas fields in each play to Pareto distributions. They "truncated" the distributions at the maximum field size and "shifted" their origins to a minimum field size; hence, the equation that describes the resulting curve is called a "Truncated Shifted Pareto" (TSP). Fitting Pareto distributions to data from discovered fields requires expert judgment. Thus, one can view the TSP function as an aid for translating expert opinion about field sizes into mathematical language.

To produce their final resource estimates, USGS assessors combined the TSP distributions with independently assessed distributions for the number of undiscovered fields in each play.

Methods for Considering Economics

As noted earlier, assessors may avoid quantitative economic analyses and the necessity for establishing explicit price and cost scenarios by defining economic boundaries qualitatively, as in the PGC's boundary of a "favorable" price/cost ratio. Problems with this approach include the difficulty of reviewing the estimates without precise boundary definitions, and the possibility—or probability—that individual assessors will apply different economic assumptions based on their own interpretations of this vague boundary condition. (At the PGC, discussions within working committees attempt to ensure that economic boundaries are applied consistently.) An advantage of this approach is its ability to recognize regional differences in production costs, infrastructure, and petroleum prices.

When an assessment includes explicit, quantitative economic analysis, analysts may choose from a number of economics "scenarios":

- They may use an actual oil and gas price forecast that varies over time, and assume drilling and other development costs are constant (adjusted only for inflation), as in the USGS/MMS assessment.
- They may present the different resource estimates for different oil prices, ranging, for example, from $15 to $30 per barrel in $5 increments.
- They may define precise physical boundary conditions tied implicitly to economic conditions: water depth limits, minimum economic field size, permeability limits, drilling depth limits, and so forth.

Analysts have shown that assessments of economically recoverable oil and gas vary considerably with world oil price. For example, the MMS evaluated the sensitivity to price of estimated recoverable oil and gas resources in offshore federal lands for its 1987 Five-Year Outer Continental Shelf Leasing Program document. This analysis found that leasable resources on the OCS varied from

9,100 million barrels of oil equivalent (Mmboe) when oil cost $17 per barrel[1] to 13,690 Mmboe when the price rose to $34 per barrel.[2] In other words, halving prices drove down recoverable resource estimates by 34 percent.

The sensitivity of estimated resource volumes to oil price depends on assumptions about costs. It is quite common to assume that oil field costs are constant (in real dollars) over time, regardless of oil prices. However, as explained earlier in this chapter, research has shown that oil-field service costs increase with oil prices (Kuuskraa *et al.*, 1987). If oil prices rise quickly, stimulating immediate increases in drilling rates, service costs will rise rapidly as well. Similarly, a rapid fall in prices, with a sharp drop in drilling and rig utilization, would drive down costs. More gradual price changes would cause less dramatic changes in costs because rig supply would have time to adjust, and shortage or surplus situations would not contribute to swings in service costs.

This added complexity in economic analysis—that oil-field service costs vary with changing prices and the rate at which prices change—creates a serious dilemma for the resource analyst. It implies that an appropriate range of scenarios for a credible quantitative analysis of recoverable resources extends to predicting not only future price levels, but also how fast they are reached and how the service industry responds. On the other hand, the variability of service costs tends to damp out the sensitivity of recoverable resources volume to price, because changes in service costs eat up some of the potential profits or losses associated with higher or lower prices (Office of Technology Assessment, 1987).

Given the importance of drilling and other costs in analyzing the economic recoverability of resources, and the potential for cost volatility, resource assessors should seek baseline economic assumptions that reflect long-term equilibrium costs, rather than costs that have been artificially inflated or deflated by rapid price changes. These costs would reflect the assumption that the supply and demand of oil-field and gas-field services are essentially in balance. This does not mean that such equilibrium costs will be constant over time, because they will change with technical advances. Unless the assessors can define clear trends in such costs, however, they probably should assume constant real costs over time. If it is desirable, assessors can examine the effect of changing costs using

---

[1]In 1987 dollars, delivered at the Gulf of Mexico.

[2] U.S. Department of the Interior, Minerals Management Service. Appendix F: Economic considerations in the 5-year outer continental shelf oil and gas leasing program, in 5-Year Outer Continental Shelf Leasing Program for January 1987-December 1991, draft.

parametric analysis. (That is, they can recalculate resource volumes at alternative costs.) Similarly, an ideal assessment would use long-term equilibrium values of oil and gas prices, avoiding values that can be sustained only for short times. Unfortunately, there is little likelihood of obtaining consensus on what these prices would be, so it makes sense for assessors to use well-known price forecasts like those from the Energy Information Administration's Energy Outlook series.

Additionally, to account for the role efficiency plays in prospect development one should choose economic and cost assumptions that reflect an "optimistic" rather than an "industry average" outlook. This choice is dictated by the nature of oil and gas development. For most prospects, there are several potential developers who, as a group, may represent an industry average in terms of drilling and service costs, required rate of return, exploration success rates, and so forth. However, it only requires one company to develop a prospect; the fact that all but one of the potential developers could not develop the prospect profitably (or would not proceed with development according to their investment decision rules) is not relevant to the prospect's inclusion in the economically recoverable resource base. Thus, the required rates of return assumed in the assessment should fall near the lower bound of all rates applicable to the group of producers capable of developing the prospects. Similarly, the assumed drilling costs should not represent the average, but should approach the costs achieved by the lowest-cost driller. It is not appropriate, however, to select the absolute lower bound (or upper bound, as appropriate) for these variables, as no individual producer is likely to be both the lowest-cost and most efficient performer in every category. Instead, one should perhaps select the lower (or upper) decile or quartile value for each variable.

A controversial aspect of selecting economic assumptions is whether or not to treat certain front-end costs as "sunk" (that is, as already spent and therefore not relevant to future investment decisions). Assessors may, in particular, treat leasing and exploration costs as sunk, on the basis that the critical economic/uneconomic decision often comes after exploration drilling, when a company has found a field and must decide whether or not to develop it. Under this assumption, assessors calculate minimum field sizes for assessment areas by comparing the value of the oil and gas found to the costs of developing the resources (i.e., the cost of building infrastructure and drilling development wells), without considering lease or exploratory drilling.

The validity of treating leasing and exploration costs as sunk depends on negative answers to two questions:

1. Might developers choose to reject prospects in advance of exploration because they fear high exploration costs?

2. If the field is abandoned before development commences, can developers recoup any of their exploration costs (i.e., is there any salvage value)?

An affirmative answer to either question implies that leasing and exploration costs cannot be ignored in deciding whether potential resources are economic.

A negative answer to the first question implies that, for all of the prospects considered, at least one company in the industry will be willing to buy a lease and conduct an exploration program. Individual geologists and companies tend to have diverging opinions about the attractiveness of most prospects; historically, company bids for individual offshore leases have varied widely. Consequently, the assumption that one company will be willing to invest in exploration is credible if the list of prospects was developed with attractiveness to industry in mind. However, this assumption would not apply to a group of prospects that represented an all-inclusive network of grid blocks over a broad area, because it is likely that some of these blocks would receive no bids. Consequently, leasing and exploration costs can be ignored in calculating economic levels of resources in a group of prospects and plays only if these costs were considered in the initial choice of prospects and plays to examine.

A negative answer to the second question implies that the current tax structure treats exploration costs nearly the same way whether a project proceeds to development or is abandoned. If the tax system treats industry costs more generously upon abandonment than upon development, the potential developer will not ignore exploration costs in his "develop or abandon" decision, and neither should the assessor, even if exploration costs were considered in selecting the list of prospects and plays. If an assessor chooses to treat leasing and exploration costs as sunk, he should justify this decision by demonstrating that any tax differences between an abandoned prospect and a developed prospect are zero or small.

RESOURCE ASSESSMENTS AND ENERGY POLICY

Resource assessments become tools for policy analysts when issues of resource scarcity, environmental sensitivity of potential resource-bearing lands, or shortfalls in resource production need to be addressed. For example:

• Government promotion of energy technologies may hinge on perceptions of the adequacy of the domestic fuels resource base. For example, perceptions of the size of the U.S. natural gas resource base will influence decisions about promoting combined cycle electricity generation fueled by natural gas.

• Proposals to stimulate drilling for oil and natural gas in the face of declining domestic production levels may hinge upon assessments of the conventional and unconventional or discovered and undiscovered part of the resource base. A conclusion that much larger volumes of natural gas reside in unconventional formations could shift policy attention to governmental assistance for production research, or to incentives aimed at tight gas or other unconventional resources. Similarly, assessment results that indicate large volumes of remaining recoverable oil in already discovered formations, with smaller volumes in the undiscovered portion of the resource base, might shift policy attention away from incentives for exploratory drilling and toward wider drilling incentives more likely to instill higher levels of infill and extension drilling.

• Policymakers evaluating proposals to sequester environmentally sensitive lands, like the Coastal Plain of the Arctic National Wildlife Refuge, will likely be strongly influenced by resource assessments covering the lands under review, the region (especially if an existing investment, like a pipeline, would lose value without continued development), and the nation.

Policymakers are interested in resources primarily because they see them as a bridge to production. That is, they interpret an oil and gas resource assessment as a means of comprehending the prospects for maintaining high levels of domestic oil and gas production in the future. Because policymakers commonly are unsophisticated about resource terminology, they are likely to interpret a sharp change in resource volumes as signalling a similar change in future production prospects. Policymakers may find it difficult to comprehend—unless told very explicitly—that a so-called national assessment of oil and gas resources actually covers only a portion of the recoverable resource base: one that may have only a modest impact on future production. This is particularly true of the oil portion of assessments: fully 70 percent of the total U.S. oil reserve additions between 1979 and 1984 came from drilling thousands of extension and infill wells in previously discovered oil fields, which are not part of the undiscovered resource base (Office of Technology Assessment, 1987). Although there is continuing debate among industry analysts about whether or not development drilling will continue to dominate reserve additions to this extent, there is no doubt that an assessment of only the undiscovered resource base cannot adequately inform policymakers about prospects for future domestic oil and gas production from all sources.

REFERENCES

Brashear, J. P., A. Becker, K. Biglarbigi, and R. M. Ray. 1989. Incentives, technology, and EOR: potential for increasing oil recovery at lower oil prices. Journal of Petroleum Technology 41(2): 164-170.

Energy, Mines and Resources Canada. 1977. Oil and Natural Gas Resources of Canada, 1976. Report 77-1. Ottawa, Canada.

Kaufman, G. M. 1965. Statistical analysis of the size distribution of oil and gas fields. Pp. 109-131 in Symposium on Petroleum Economics and Evaluation. Richardson, Texas: Society of Petroleum Engineers.

Kaufman, G. M., Y. Balcer, and D. Kruyt. 1975. A probabilistic model of oil and gas discovery. Pp. 113-142 in Studies in Geology No. 1: Methods of Estimating the Volume of Undiscovered Oil and Gas Resources. Tulsa, Oklahoma: American Association of Petroleum Geologists.

Kuuskraa, V. A., F. Morra Jr., and M. L. Godec. 1987. Importance of cost-price relationships for least-cost oil and gas reserves. Society of Petroleum Engineers Paper No. 16289. Presentation to the Society of Petroleum Engineers Hydrocarbon Economics and Evaluation Symposium, March 1987, Dallas, Texas.

Lee, P. J., and P. C. C. Wang. 1983a. Probabilistic formulation of a method for the evaluation of petroleum resources. Mathematical Geology 15: 163-181.

Lee, P. J., and P. C. C. Wang. 1983b. Conditional analysis for petroleum resource evaluations. Mathematical Geology 15: 353-363.

Lee, P. J. and P. C. C. Wang. 1985. Prediction of oil or gas pool sizes when the discovery record is available. Mathematical Geology 17: 95-113.

Office of Technology Assessment. 1987. U.S. Oil Production: The Effect of Low Oil Prices—Special Report. OTA-E-348. Washington, D.C.: U.S. Congress.

Oil and Gas Journal. 1989. U.S. fields with reserves exceeding 100 million barrels. January 30, pp. 69-70.

Podruski, J. A., J. E. Barclay, A. P. Hamblin, P. J. Lee, K. G. Osadetz, R. M. Procter, and G. C. Taylor. 1988. Conventional Oil Resources of Western Canada. Paper 87-26. Ottawa, Canada.

Roy, K. J. 1979. Hydrocarbon assessment using subjective probability and Monte Carlo methods. Pp. 279-290 in First IIASA Conference on Methods and Models for Assessing Energy Resources, M. Grenon, ed. New York: Pergamon Press.

Roy, K. J., R. M. Procter, and R. G. McCrossan. 1975. Hydrocarbon assessment using subjective probability: probability methods in oil exploration. Pp. 56-60 in Probability Methods in Oil Exploration: American Association of Petroleum Geologists Research Symposium Notes, J. C. Davis, J. H. Doveton, and J. W. Harbaugh, eds. Tulsa, Oklahoma: American Association of Petroleum Geologists.

U.S. Department of the Interior, Geological Survey and Minerals Management Service. 1989. Estimates of Undiscovered Conventional Oil and Gas Resources in the United States—A Part of the Nation's Energy Endowment. Washington, D.C.: U.S. Government Printing Office.

U.S. Geological Survey. 1975. Geological Estimates of Undiscovered Recoverable Oil and Gas Resources in the United States. Circular 725. Washington, D.C.: Department of the Interior.

U.S. Geological Survey. 1980. Future Supply of Oil and Gas from the Permian Basin of West Texas and Southeastern New Mexico. Circular 828. Washington, D.C.: U.S. Department of Interior.

U.S. Geological Survey. 1981. Estimates of Undiscovered Recoverable Conventional Resources of Oil and Gas in the United States. Circular 860. Washington, D.C.: Department of the Interior.

White, D. A. and H. M. Gehman. 1979. Methods of estimating oil and gas resources. American Association of Petroleum Geologists, Bulletin 63: 2183-2192.

3

AN EVALUATION OF THE DEPARTMENT OF THE INTERIOR'S 1989 RESOURCE ASSESSMENT

As Chapter 1 described, the Department of the Interior's 1989 resource assessment caused concern among some petroleum industry representatives. The assessment projected significantly lower undiscovered resource values than two prior DOI assessments, issued in 1975 and 1981. Certainly, part of the decrease was due to discoveries made between assessments. Once an exploratory drill uncovers a petroleum field, the field moves from the "undiscovered" category to the category of "identified" resources awaiting extraction. But, more important, the estimates from the different years may have varied because of limitations of the assessment methodology.

This chapter presents the Committee on Undiscovered Oil and Gas Resources' evaluation of the methodology used in the 1989 assessment. It begins with a brief overview of the assessment boundaries, important for understanding the assessment results. Then, it moves to a two-part evaluation of the assessment methodology: the first part covers the USGS; the second part covers the MMS. (Recall from Chapter 1 that the USGS inventoried resources for oil-producing regions onshore and in state waters, while the MMS assessed resources beneath the Outer Continental Shelf.)

ASSESSMENT BOUNDARIES

Understanding the assessment boundaries—which oil and gas resources the assessment includes and which it excludes—is crucial for policymakers who are formulating energy strategy and attempting to gain a comprehensive view of future oil and gas resources available for development. This section evaluates three key parameters related to the assessment boundaries: the definitions of conventional and unconventional resources, the economic assumptions, and the reserve growth of existing petroleum reservoirs.

Conventional/Unconventional Boundary

One controversial aspect of the DOI's assessment is its limitation to "conventional" resources. As Chapter 1 explained, industry groups—most notably the Potential Gas Committee—do not share the DOI's delineation between conventional and unconventional resources and include in their own assessments certain resources that the DOI defined as outside of the assessment boundaries.

The DOI assessment defined conventional resources as "crude oil, natural gas, and natural gas liquids that exist in conventional reservoirs or in a fluid state amenable to extraction techniques employed in traditional development practices" (U.S. Department of the Interior, 1989). Further, the assessment specified that these resources must exist in "discrete accumulations." The USGS, which was responsible for the onshore assessment that covered the bulk of unconventional resources, acknowledged that the boundary between conventional and unconventional resources is hazy in many situations.

The DOI's reasons for setting aside unconventional reservoirs from analysis were: (1) they are mostly discovered and their geographic and stratigraphic distributions are largely known; and (2) in-place volumes are large and recoverability is uncertain because of technical and economic factors (U.S. Department of the Interior, 1989). Further, the DOI judged that the geologic data and assessment methods available for evaluating unconventional resource occurrences were inadequate.

In the committee's opinion, the DOI's decision to exclude unconventional *oil* resources from its assessment is justifiable. Except for thermally stimulated heavy oil recovered primarily in California, unconventional oil deposits, like those from tar sand and oil shale, have contributed little to domestic oil production. Indeed, current and future production costs of these resources will be higher than costs of conventional resources.

On the other hand, the DOI's decision to exclude unconventional *natural gas* resources presents problems. Natural gas deposits in low-permeability sandstones, fractured shale, and coal beds—sources the DOI labelled "unconventional"—are making an important contribution to current production, variously estimated at about 1.5 to 2 trillion cubic feet (Tcf) per year. Much of this volume is produced from low-permeability gas-bearing sandstones and is primarily the result of the accelerated drilling and development that occurred between 1977 and 1982, especially between 1979 and 1982 under incentives from the Natural Gas Policy Act. Production of coal-bed methane and natural gas from fractured shale have been increasing in recent years, with the former supported by increased understanding of the resource's distribution, new production technologies, and tax incentives. The recent (December, 1989) extension of coal-bed methane tax credits and the reimplementation of tax credits for tight-gas reservoirs for two years will assist further development of these resources. Production of all major forms of unconventional natural gas has been aided by extraction research funded by the Gas Research Institute and the U.S. Department of Energy.

While acknowledging some of the special characteristics of unconventional natural gas accumulations, the committee concludes that it is both feasible and appropriate to include these resources in a national resource assessment.[1] Of course, the estimates of unconventional resources should be provided separately from estimates of conventional resources. Industry decisions about whether to produce unconventional resources are guided by a unique set of technical characteristics and economic incentives (in the form of tax credits). Thus, estimates of unconventional resources are most useful when they are provided separately, because future production of these resources may depend upon economic and technical conditions different from the conditions that influence the production of conventional resources.

In-Place Resources: Recoverable/Unrecoverable Boundary

As discussed in Chapter 2, the classification of resources as "recoverable" or "unrecoverable" with current technology is at best a snapshot in time. Advances in reservoir characterization, drilling, and completion continually increase the percentage of a reservoir's total petroleum supply that is recovered. Yet, the USGS/MMS assessment (like most other assessments) included no estimates of in-place resources, only estimates of resources recoverable with

[1]In this chapter, the committee's conclusions and recommendations are underlined.

current technology. Estimates of in-place resources are important in judging the potential for new recovery technology to increase the producible petroleum supply.

Because the assessment was limited to recoverable resources, on average only about one-third of discovered oil and 60 to 85 percent of discovered natural gas was incorporated in the models used to estimate undiscovered resource volumes. Excluding consideration of in-place resources in these models imposes a technological overlay too early in the assessment process. A more thorough evaluation would first estimate each play's in-place resources and then multiply these estimates by recovery efficiencies to estimate technically recoverable resources. Using estimates of in-place resources in combination with recovery efficiencies would allow the estimates to be updated more easily to account for advances in recovery technology.

The committee recommends that in future assessments, the USGS and MMS develop methods to estimate in-place resources in each play. The agencies should base their estimates of technically recoverable resources on in-place resources, by applying recovery factors to in-place resource estimates.

Economic Boundaries

The assessment's economic boundaries hinged on the computation of a "minimum economic field size" (MEFS): the smallest field that assessors determined can be developed profitably. Calculating the MEFS requires a myriad of economic assumptions. For the 1989 assessment, the fundamental assumptions were:

- The MEFS must yield a net average after-tax return of 8 percent on the development investment.
- The absolute price of natural gas would not exceed 75 percent of the btu-equivalent price of crude oil.
- The 1987 oil price was $18.00 per barrel.
- The 1987 natural gas price was $1.80 per million cubic feet (Mcf).
- Real oil prices would decline at an annual rate of 3 percent between 1987 and 1989 and then increase 4 percent (with a range of plus or minus 1 percent) per year beginning in 1990.
- Real gas prices would decline 2 percent annually between 1987 and 1989 and then increase 5.5 percent (with a range of plus or minus 1 percent) per year beginning in 1990.

• The future costs for reservoir development and transportation facilities would remain fixed at 1985-1986 levels.

With these assumptions, DOI analysts calculated the MEFS on the basis of a 1987 field discovery and a decision to develop the field beginning in that year.

The USGS and the MMS used the same assumptions in determining the MEFS. However, because onshore and offshore assessments differ in their overall approach and methodology, the USGS and MMS used different boundaries to separate the economically recoverable resources from the subeconomic resources.

USGS Economic Boundaries

In its analysis of the lower 48 states onshore and the offshore state waters, the USGS defined plays using field sizes down to 1 million barrels of crude oil or the energy-equivalent volume of natural gas (denoted as Mmboe; 1 Mmboe is equal to 6 billion cubic feet (Bcf) of gas). The 1 Mmboe volume was used to truncate the geologically assessed distribution of undiscovered fields and to provide a lower boundary for the play assessment methodology (Attanasi, 1988). In part, this truncation was a function of the minimum field size included in the national data base used as input for the assessment. For the lower 48 states onshore, the USGS found that virtually all fields of 1 Mmboe or larger were economic to develop, based on depth and field size. However, for the assessment of fields smaller than 1 Mmboe, the USGS calculated the MEFS by province or region and then determined what fraction of these small fields was economic to develop.

The calculation of the MEFS was done by discounted cash-flow analysis using assumed future oil and gas prices, inflation rates, rates of return, development costs, and timing of development (U.S. Department of the Interior, 1989). Estimates of development costs accounted for drilling depth, especially for small fields, and water depth for state waters (Attanasi, 1988).

Overall, only 1 in about 2,100 fields containing more than 1 Mmboe was found to be uneconomic to develop in the onshore lower 48 states. Fifty-two percent of 50,241 undiscovered oil fields and 47 percent of 27,014 nonassociated gas fields smaller than 1 Mmboe were considered economic to develop (Attanasi, 1988). The committee concludes that since tens of thousands of undiscovered small fields were considered economical to develop, the chosen minimum size cutoff of 1 Mmboe was too high for the overall assessment.

It is important to note that for both Alaska and state waters in the Gulf and Pacific, where development is more expensive, the USGS used a higher MEFS. For the Gulf and Pacific, the USGS assumed a minimum economically developable crude oil field size of 2 Mmboe. For Alaska, economic constraints were even greater.

For the overall assessment, in moving from *technically* recoverable to *economically* recoverable resources, applying the MEFS and the economic parameters (most notably the $18.00 per barrel crude oil price and the $1.80 per Mcf natural gas price) reduced the undiscovered resources onshore in the lower 48 states by only a small amount. Nationally, for crude oil, there was a 20 percent reduction (6.7 billion barrels) between onshore mean undiscovered recoverable resources and onshore mean undiscovered economically recoverable resources. Of that 6.7 billion barrels, however, 5.3 billion barrels were in the Alaskan region. For natural gas, there was a 26 percent reduction (65.3 Tcf) between onshore mean undiscovered recoverable resources and onshore mean undiscovered economically recoverable resources. Of that 65.3 Tcf, 56.7 Tcf were in the Alaskan region. Thus, for the onshore lower 48 states, the economic overlay produced only a small constraint to the undiscovered resource volumes.

The important implication of these results—an implication policymakers must recognize—is that higher oil prices will NOT by themselves transform large volumes of undiscovered resources from technically recoverable to economically recoverable, except in Alaska. Assuming that the DOI assessment is correct, a higher oil price may speed recovery of these resources by offering a higher profit, but it will not lead to much higher total (ultimate) recovery unless the price increases stimulate development of technological advances that expand the boundaries of technical recoverability (for example, to deeper waters).

We note, however, that the estimation of technically recoverable resources in the USGS assessment used an economic screen (for example, a minimum field size). The use of such a screen is contrary to the meaning of the term "technically recoverable" and implies that some high-cost but recoverable resources were left out of the assessment.

MMS Economic Boundaries

For its resource evaluation, the MMS divided the OCS into "planning areas." Each planning area consisted of one or more basins. Each basin contained one or more groups of identified or postulated reservoirs that the MMS called "plays," although these plays were actually summations of prospects and not true, geologically defined plays.

The MMS used a mathematical computer model, PRESTO (Probabilistic Resource Estimates—Offshore), to produce its final resource assessments. The PRESTO model produces a range of resource estimates with a corresponding estimate of the probability of occurrence. It simulates drilling of the modeled prospect and hydrocarbon discovery. MMS analysts derived probability distributions of resource volumes from multiple runs of the model. PRESTO determines the decision to develop, which would involve platform and production well installation, by comparing the prospect resources with a MEFS. For the 1989 assessment, the MEFS was determined outside the PRESTO model with a model called MONTCAR, a discounted cash-flow analysis program that calculates the volume of resources needed to balance various operating costs. In calculating operating costs, MONTCAR considers water depths, drilling depths, distance from shore, and other operating conditions for each prospect (Minerals Management Service, 1985). This results in a range of costs for different areas.

The MMS's calculation of the MEFS was prospect specific. It did not incorporate costs associated with production infrastructure, such as pipelines and onshore processing facilities, that might be shared between multiple discoveries. If, on a specific trial of the PRESTO model, the computed resources exceeded the MEFS, the model stored the results for developing the final range of outcomes. If the resources for that prospect were less than the MEFS, the model set the resource volumes to zero for that trial. The model also calculated minimum basin reserves (MBR) and minimum area reserves (MAR) to determine if the aggregate estimated undiscovered resources were adequate to justify required transportation and plant facilities (U.S. Department of the Interior, 1988).

The MEFS for the base-case economic scenario ($18.00 per barrel of oil and $1.80 per Mcf of gas) ranged from 3 Mmboe in the Gulf of Mexico and the Pacific, to 5 Mmboe in the Atlantic, to between 44 and 299 Mmbo in Alaska (depending on location). The maximum MEFS ranged from 190 Mmboe in the Pacific to 690 Mmboe in the Gulf of Mexico, and from 300 Mmboe in the Bering Sea of Alaska to 1,000 Mmboe in the Atlantic. MAR volumes ranged from zero in the Gulf and Pacific (that is, infrastructure is already available in these areas), to 120 Mmboe in the Atlantic, to between 77 and 810 Mmbo in Alaska (U.S. Department of the Interior, 1988). For regions with an established producing infrastructure, the MEFS and MAR had little effect on development of the undiscovered resource. In more remote or severe operating portions of producing regions, or in regions where no production has been established, economic boundaries were present. Such boundaries were significant where costs of building infrastructure are high, as in remote, deep-water areas and other places where operating conditions are difficult. Economic boundaries were moderate for the Atlantic region but were much more significant for Alaskan waters because the cost of building infrastructure there is so high.

Inferred Reserves and Reserve Growth

Though the DOI assessment focused on petroleum from undiscovered reservoirs, it also included a separately reported estimate of inferred reserves (the postulated incremental but unknown volume estimated to be recoverable from known reservoirs). Inferred reserves are an important component of the resource base. The DOI assessment, for example, estimated that inferred oil reserves are 63 percent as large as undiscovered oil reserves, based on mean values. Therefore, the accuracy of the method used to estimate inferred reserves is an important issue.

The DOI estimates of inferred reserves were based on a statistical time series of ultimate recovery by year of field discovery that ended in 1979. This time series was compiled and published by the American Petroleum Institute (API) and the American Gas Association (AGA). It captures historical and traditional reserve growth sources, such as extensions and new pools. The time series data reflect drilling experience probably no more recent than 1977. The period since 1977 has seen a substantially increased understanding of reservoir heterogeneity, a greater understanding of the consequences of sweep efficiency in waterfloods, and the potential for strategically targeting infill drilling and recompletions. As a result, the data through 1977 on reported inferred reserves do not reflect the increased knowledge of reservoirs gained in more than ten years of drilling. At the time of the assessment, there was no way to avoid this shortcoming. However, since the assessment was completed, the Energy Information Administration (EIA) has prepared an evaluation of oil and gas reserves by year of discovery that could help document the increasing efficiency in the conversion of discovered resources to reserves (Energy Information Administration, 1990). The estimated ultimate recovery (EUR) of oil and natural gas was compiled for six dates between 1977 and 1988 by year of discovery in groups of five years each. These new data can be used, for example, to show that for Railroad Commission District 8 in Texas—a leading oil-producing district—the annual average increase in EUR for 1977-1988 was 1.5

BOX 3.1 The USGS'S Legislative Mandate

The U.S. Geological Survey has deep roots in the monitoring of the nation's natural resources. In 1879, as the nation pushed its boundaries west, Congress, at the recommendation of the National Academy of Sciences, created the USGS to map the new territories. Congress mandated in the Organic Act of 1879 that the USGS oversee "classification of public lands and examination of the geological structure, mineral resources and products of the national domain." In response, the new agency dispatched teams of scientists on horseback to document the west's natural resources.

Since 1879, the USGS's mission to research energy resources in various parts of the nation has evolved through a patchwork of federal laws, including:

- Public Law 29, passed in 1935, and the Appropriations Act of Fiscal Year 1959 mandated that the USGS extend its mineral resource investigations to Puerto Rico, Antarctica, and the Trust Territory of the Pacific Islands.
- The Wilderness Act of 1964 requested that the USGS assess mineral resources of areas proposed or established as wilderness sites.
- The Alaska National Interest Lands Conservation Act of 1980 required that the USGS assess the oil and gas potential for federal lands in Alaska.

Although the USGS is responsible for developing and disseminating the geologic information needed to help formulate policy and ensure the wise development of the nation's energy resources, there is no provision in the federal statutes for a comprehensive national program to inventory oil and gas (National Research Council, 1988).

percent for fields discovered between 1920 and 1934. This was an increase from a rate of 1.0 percent in the period 1971-1977 for those same discoveries.

The DOI should determine if the EIA's new data offer a method of updating previously inadequate recognition of reserve growth potential of known heterogeneous reservoirs. The DOI should also determine whether use of these new data are suitable for, and will continue to be available for, an improved inferred reserve assessment methodology. If these data are not appropriate or will not continue to be updated, other methods to define reserve growth potential should be developed.

DETAILED EVALUATION OF USGS ASSESSMENT METHODS

The USGS has the responsibility to assess the oil and gas resources of the onshore portions of the United States (see Box 3.1). Its jurisdiction includes the submerged lands contiguous to the coastal states to a distance of three miles from their coasts. This section evaluates the quality of the USGS assessment for these areas.

Organization and Staff

Oil and gas resource assessments carried out by the USGS are the responsibility of the Office of Energy and Marine Geology in the Geologic Division. Two branches, headquartered in Denver, provide the staff.

For the 1989 assessment, the USGS divided the nation into 9 regions and subdivided the regions into 80 provinces (see Figure 3.1). The provinces were assigned to 42 geologists from either the Petroleum Geology Branch or the Sedimentary Processes Branch. Following procedures set forth by an assessment coordinating committee, each geologist carried out the assessment for his or her province. Each of the 42 province geologists was responsible for preparing a report on the geology and oil and gas plays in each province.
(The committee found, however, that some of these reports were not prepared and completed as open-file reports until after the assessment was finished.) Work was reviewed by the assessment coordinating committee, which made the final decisions about whether to use or modify the data provided by province geologists. The USGS reported that each of the 42 province geologists spent not more than half time on the assessment over a two-year period. The project

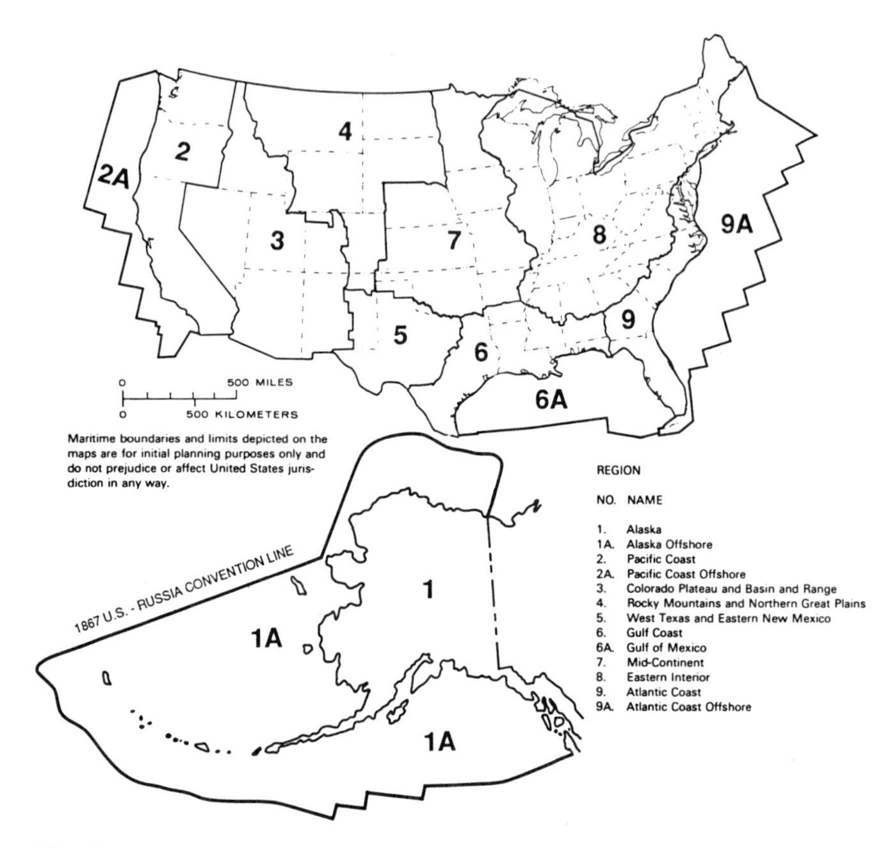

FIGURE 3.1 The nine regions into which the DOI divided the nation for purposes of assessing undiscovered oil and gas.

leader was the only USGS staff member who spent almost full time on the assessment.

The committee found that experienced geoscience personnel were distributed unevenly among the provinces, and that their distribution did not correlate with the importance of the provinces as producers of oil and gas or their potential for containing undiscovered resources. Current research interests of participating geologists rather than the resource assessment appeared to dictate staff allocations. This resulted in a concentration of attention on the Rocky Mountains, California, and Alaska and a comparatively low level of effort in the Gulf Coast, Midcontinent, and Illinois Basin areas. In the lower 48 states, the Gulf of Mexico region had the highest level of undiscovered recoverable natural gas and the second-highest level of undiscovered recoverable oil, yet it received a more limited allocation of staff than other regions. For this region, at least one extremely large play was found to be exceptionally diverse in its geologic character. The possibility exists that additional resource potential was masked because of a high degree of aggregation of reservoir types within this play. This potential is one that could have been tested but was not given the necessary allocation of effort within the assessment.

Members of the assessment coordinating committee were veterans of the previous USGS assessment and each had several years of experience in assessment methodology and analysis. In addition to reviewing the work of the province geologists, this group was responsible for determining the use of the data they provided and for applying the analytical assumptions and techniques that produced the results. The assessment coordinating committee was assembled solely for this particular project; a permanent resource assessment group no longer exists in the USGS.

In contrast to the considerable level of experience in the assessment coordinating committee, the amount of experience in play analysis and resource assessment among the province geologists varied greatly. In at least one case, one of the two geologists assigned to a major province had limited experience with the geology and oil and gas production patterns within that province. In another case, an experienced geologist was assigned to a major province, but the province was large and the assessment effort could have benefitted from a greater disaggregation of plays. And, though analysts may have been experts in the general geology of the provinces for which they were responsible, many had little or no experience in resource assessment or the methodology employed. This necessitated training by the assessment coordinating committee for many of the province geologists. The committee concluded that because the play analysis procedure used for the assessment depends on expert judgment and knowledge of geologic factors affecting oil and gas accumulation, the lack of experience could be a significant factor in the accuracy of the results.

Data Base

Sources

Because no publicly accessible, comprehensive data base on United States oil and gas fields exists, the USGS used a wide variety of data sources for its assessment. As discussed in Chapter 2, several data bases, both public and private, cover one or more elements of in-place petroleum. (TORIS is one example.) However, each of these data bases is incomplete in coverage either in the total petroleum content of each reservoir in the data base or in the coverage of all discovered reservoirs.

The committee found that the USGS assessment was limited in particular by a lack of seismic data for the lower 48 states. While recognizing the enormity of the task of collecting such data, some committee members questioned whether an adequate assessment could be completed without such a data base. Without seismic data covering the lower 48 states, the USGS must depend on gathering as much geologic and related petroleum information as possible from all available sources. The data gathering task is difficult, because there are three categories of land ownership in the onshore: federal, state, and private. The number of industrial participants in onshore petroleum exploration and development is about two orders of magnitude larger than the number engaged in similar offshore activities. Onshore seismic data commonly are proprietary to the company contracting for the work or to a seismic contractor doing the work on a speculative basis. Such data often are not available to the USGS or, if they are available, are expensive. (This situation contrasts with offshore, where the federal government owns all the resources and the MMS has access to seismic data at only the cost of reproduction.) In part because of the difficulty of obtaining data, it appears that the USGS used mostly its own information, from both published and unpublished sources, for the 1989 assessment.

The committee recommends that in future assessments, the USGS make a more consistent effort to seek out both published and unpublished information from state geological surveys, state regulatory agencies, and private-sector sources. The USGS should also alert assessment users to the enormous variety of the data on which it bases its final estimates.

The province or basin reports prepared by the province geologists represent a synthesis of the data utilized in each area. The detail and degree of orientation toward petroleum assessment needs of these reports varied greatly. This reflected the variety of backgrounds and research interests of the preparers.

While the USGS plans to place all of these reports in its open-file series, only 24 were listed as complete, with an additional 16 in review or revision, at the time of this committee's review.

Association of American State Geologists' Review of USGS Data

In 1988, the Secretary of the Interior requested—in response to the natural gas industry's questions about the assessment—that the Association of American State Geologists (AASG) review the geologic information used in the assessment (Association of American State Geologists, 1988). The AASG's review was intended to be complementary to this report. The AASG solicited comments from the geologic community (including geologists from academia, state agencies, and industry) through a series of regional workshops. At each workshop, representatives of the USGS and/or the MMS presented the geologic information they used in the assessment process for the region in question. Workshop participants were asked to provide written comments on the following topics:

1. their overall impression of the adequacy of the geologic information used in defining plays and in carrying out play analysis;
2. whether they knew of any geologic information from state surveys or other sources that the USGS did not use, but that would have significantly improved the quality of the assessment; and
3. whether they noted any deficiencies or unanswered questions in areas in which they had particular expertise.

The findings from the AASG report mirrored some of this committee's observations. While AASG reviewers noted the extensive amount of information available to the MMS, they commented on the lack of seismic data available to the USGS and on the variability in quantity and quality of other USGS data. The AASG report cited several specific examples of information that was readily available and pertinent to the analysis of a province, but was not drawn upon.

Data Auditing

Good statistical practice in resource appraisal begins with an audit of the data used in the assessment: data describing predrilling exploration, well outcomes, and the depositional environment of individual prospects. A data audit is a framework for: (1) evaluating the data's accuracy and completeness; (2) identifying areas where the data require improvements; and (3) providing explicit measures of the data's quality to assessment users. The USGS currently lacks formal data audit procedures.

The committee recommends that the USGS design an audit to assess the quality of exploration data employed in its resource assessments. The USGS could start by instituting a formal audit of the U.S. oil and gas field data files used in the 1989 assessment. These data files include Petroleum Information Corporation's PDS oil and gas field files, NRG Associates' Significant Oil and Gas Fields of the United States files, and state geological survey and state oil and gas commission field files. Such an audit is well within the scope of the resource surveys undertaken by the USGS and would enhance the credibility of future resource assessments.

Play Analysis

This assessment was the first in which the USGS used a play-analysis approach on a national scale. The USGS identified 250 plays containing fields with more than 1 million barrels of oil or 6 billion cubic feet of natural gas (U.S. Department of the Interior, 1989). These 250 plays formed the foundation for the assessment. The definition of plays therefore is a critical component of the overall assessment process; it is the starting point for the prediction of undiscovered resource volumes.

Background

A play-based resource assessment method attempts to group into "plays" known reservoirs or fields that have common characteristics. It then predicts the remaining undiscovered petroleum accumulations within each group. Thus, play analysis provides a framework within which predictions about undiscovered resources are closely tied to knowledge about the distribution of already-discovered resources within that play.

White defined a play as "a group of geologically similar prospects having basically the same source-reservoir-trap controls of oil and gas" (White, 1980)—a definition suggesting reservoir character as a key element in play definition. Fisher and Galloway found, on the basis of analysis of all the major oil reservoirs in Texas, that "the most unifying, first-order character of play definition is the genetic origin of the reservoir," by which they meant the commonality of the depositional system responsible for laying down reservoir rocks (Fisher *et al.*, 1984). Thus, *reservoir origin* is widely regarded as an important parameter in play definition (Tyler *et al.*, 1984). According to this view, depositional systems affect the external and internal geometries of reservoir rocks—critical determinants, respectively, of field size and field distribution within a basin. The depositional systems, therefore, affect the overall architecture of pay and nonpay zones.

Plays must be drawn up carefully and must be as internally consistent as possible in terms of their geology and their reservoir properties. The mixing of dissimilar reservoirs in fields making up a play may lead to inaccurate statistical characterization of the play's maturity. Inaccurate statistics, in turn, may lead to an inaccurate assessment of the play's potential to contain undiscovered resources. Canadian assessors have recognized that poor play definition or mixing of plays may distort statistical analysis (Podruski *et al.*, 1988).

Sources of Information About Plays

The USGS derived its 250 plays primarily from the Significant Oil and Gas Fields of the United States Data Base, a proprietary data base produced by NRG Associates. The version of this data base the USGS used contained about 370 "clusters," each composed of a set of fields whose volumes of ultimately recoverable petroleum were at least 1 Mmboe. In this data base, a major reservoir is located in only one cluster, but a field containing more than one reservoir may occur in more than one cluster. NRG's definition of a cluster was not necessarily designed to conform to an appropriate play definition. Nevertheless, the NRG data base provided a reasonably comprehensive national information source that the USGS could use to develop size distributions for discovered petroleum at the time of the assessment. In addition, the data base was advantageous because it was in a machine-processable format.

The USGS interpreted some clusters as plays and aggregated others to form plays. USGS geologists then provided judgments on play and accumulation attributes, including the existence of a hydrocarbon source, favorable timing of hydrocarbon migration, potential migration pathways, and the existence of potential reservoir rock facies (Crovelli *et al.*, 1988).

The USGS's Application of Play Analysis

Although the committee regards the USGS's use of play analysis as a substantial improvement over previous methods, the committee is concerned about how accurately the USGS applied the play analysis method.

Play definition: Applying the play concept requires the identification of many factors, including depositional systems, reservoir structure, origin and migration of petroleum, and trapping style. The scales of the major components that form reservoirs (e.g., the thickness and length of a barrier bar or the areal extent of a delta lobe) are related to the field-size distribution within a basin. Lithology (in the gross sense of the major categories of carbonates versus sandstone) commonly can be a quick first check on any dichotomies in a play composed of smaller groups of reservoirs that might otherwise be assumed to be depositionally similar. Structural geological considerations may override those of depositional systems (where judged appropriate), but purely geographic groupings generally should be avoided.

The committee found that in some cases, USGS assessors created excessively large and lithologically diverse plays from clusters within the NRG data base. For example, in the Gulf Coast province, the USGS created extensive plays from vertical (and horizontal, in at least one major play) aggregations containing such a diversity of Tertiary depositional systems that the committee questioned whether they met the criteria of a play. In East Texas, the USGS combined both carbonate and sandstone clusters in the same play. The committee judged that this tendency to combine diverse aggregations in a single play seemed most significant in the Permian Basin. The mixing of dissimilar reservoirs in fields making up a play, as may have occurred in the Gulf Coast, East Texas, and Permian Basin areas, may cause either overestimated or underestimated resource volumes. Play mixing could *inflate* estimated resource volumes by masking declines in discovery sizes that normally occur as exploration progresses in the area encompassing the play. Play mixing could *decrease* resource estimates by biasing statistical computations so that the play appears mature, with many fields discovered and few remaining undiscovered fields (see the discussion below). Whether play mixing results in overstated or understated resource estimates is critically dependent on the time pattern in which the discovery wells are drilled within the mixed plays. (See the discussion of staggered mixing of plays in Adelman *et al.*, 1983.)

To determine how play mixing might have affected the USGS's resource estimates, the committee analyzed in detail ten plays in the Permian Basin, where we judged that the tendency to combine diverse aggregations in a single play seemed most significant and most amenable to evaluation. (Appendix C provides a detailed discussion of our analysis.) We analyzed plays containing mixed sandstone and carbonate lithologies to determine the environments of deposition of the various lithic types. We found that where plays were almost one-third sandstone or carbonate, with the balance being the other lithic type, they consisted of platform carbonate and submarine fan/canyon-fill sandstone.

Our analysis showed that partitioning the field discovery history of plays with mixed depositional systems into thirds, in the same manner as the USGS, using the same dates, yields a distribution that resembles the histogram of a play within which many fields have already been discovered, rather than a play with a greater proportion of undiscovered fields. It is likely that the mixing of dissimilar depositional systems in a play, with different characteristic size distributions for the reservoirs in each system within that play, can create a play that appears excessively mature in its discovery history.

The detailed analysis described in Appendix C suggests that the formulation of plays by aggregating clusters that include different depositional systems will underestimate the volume of undiscovered resources. Assuming the that the field-size distribution within a play bears some relationship to the depositional systems within that play, the size frequency distribution of a mixed play, being more varied, will appear more complete. This will lead to the misinterpretation that there are fewer undiscovered fields.

The committee cannot determine the degree to which the assessment is biased by the aggregation of diverse depositional systems into plays. Concerns about the makeup of plays do not extend equally to all regions and provinces assessed, and for many areas these concerns do not exist. However, we believe that such aggregations could produce important miscalculations in some circumstances. We are concerned that the USGS has performed no tests to evaluate this possibility.

The committee recommends that the USGS analyze and refine play content before the next assessment to determine the impact of play formulation on resource volumes. The USGS should also make statistical testing of play formulation a required part of the methodology for future assessments so that play mixing does not limit the reliability of the estimates.

Conceptual Plays: A further constraint on the current assessment is its general lack of treatment of "conceptual" plays: those that do not contain discoveries or reserves but that geological analysis indicates may exist (Podruski *et al.*, 1988). Conceptual plays cannot be evaluated with the same analytical

<table>
<tr><th colspan="3">Attribute</th><th colspan="7">Probability Favorable or Present</th></tr>
<tr><td rowspan="5">Play Attributes</td><td colspan="2">Hydrocarbon Source (S)</td><td colspan="7"></td></tr>
<tr><td colspan="2">Timing (T)</td><td colspan="7"></td></tr>
<tr><td colspan="2">Migration (M)</td><td colspan="7"></td></tr>
<tr><td colspan="2">Potential Reservoir Facies (F)</td><td colspan="7"></td></tr>
<tr><td colspan="2">Marginal Play Probability
$S \times T \times M \times R = MP$</td><td colspan="7"></td></tr>
<tr><td>Accumulation Attribute</td><td colspan="2">Conditional Probability of at least one undisc. accumulation in play

Minimum accumulation size assessed:
_____ $\times 10^6$ BBL; _____ $\times 10^9$ CF</td><td colspan="7"></td></tr>
<tr><td rowspan="10">Hydrocarbon Accumulation Parameters (Undisc. accum's)</td><td rowspan="3">Reservoir Lithology</td><td>Sandstone</td><td colspan="7"></td></tr>
<tr><td>Carbonate</td><td colspan="7"></td></tr>
<tr><td>Other</td><td colspan="7"></td></tr>
<tr><td rowspan="2">Hydrocarbon type</td><td>Gas</td><td colspan="7"></td></tr>
<tr><td>Oil</td><td colspan="7"></td></tr>
<tr><td colspan="2">Fractiles / Attribute</td><td>100</td><td>95</td><td>75</td><td>50</td><td>25</td><td>5</td><td>0</td></tr>
<tr><td rowspan="2">Accumulation Size</td><td>Oil (10^6 BBL)</td><td></td><td></td><td></td><td></td><td></td><td></td><td></td></tr>
<tr><td>Gas (10^9 CF)</td><td></td><td></td><td></td><td></td><td></td><td></td><td></td></tr>
<tr><td rowspan="2">Reservoir Depth ($\times 10^3$ Ft.)</td><td>Oil</td><td></td><td></td><td></td><td></td><td></td><td></td><td></td></tr>
<tr><td>NA Gas</td><td></td><td></td><td></td><td></td><td></td><td></td><td></td></tr>
<tr><td colspan="3">Conditional No. of accumulations</td><td></td><td></td><td></td><td></td><td></td><td></td><td></td></tr>
</table>

FIGURE 3.2 The appraisal form that USGS assessors used to record geologic information about each play.

procedure used to assess plays within existing fields; the approach to assessing these plays would be much more subjective. Results would have to be reported separately, recognizing the potential for major uncertainties related to the adequacy of source rocks, reservoir rocks, and trapping mechanisms.

Conceptual plays may be especially important for assessing natural gas supplies, because natural gas exploration is less mature than oil exploration and because future natural gas discoveries will likely be found at deeper stratigraphic intervals than oil. Information was available from industry and state sources that would have supported a conceptual gas play for Arbuckle strata in the Anadarko basin of eastern Oklahoma and western Arkansas. Recent exploration in Arbuckle rocks in that basin has turned up gas reserves in excess of 600 Bcf. A similar conceptual play for Arbuckle strata in the Anadarko basin also would have been possible. That interval in the Anadarko basin remains to be tested.

The committee determined that the USGS did not provide adequately for unknown or poorly known plays in the current assessment. This omission is a point that must be addressed in future assessments.

Quantification of Expert Judgment and Statistical Methods

Background: The Appraisal Form

The USGS used an appraisal form to record information about each defined play (see Figure 3.2). The form included spaces for geologic risk factors, a distribution for the number of petroleum fields, separate field-size distributions for crude oil and natural gas, and gas/oil probabilities. The distributions that represent the field size and number of fields were specified in terms of an absolute maximum and minimum together with fifth, twenty-fifth, fiftieth, seventy-fifth, and ninety-fifth percentiles. These distributions were subjectively assigned by consensus, based on the discovery history and geological information for the play. Piece-wise linear cumulative distribution functions were fitted between the assigned maxima and minima, passing through the assigned percentiles. For each of these fitted distributions, a computer program calculated a mean value and a variance. Only the means and variances of the assigned distributions for each play were used for the resource appraisal.

For portions of Alaska's North Slope, where extensive seismic information is available, the USGS used a procedure based on identified individual prospects, similar to the MMS's approach for well-mapped federal offshore tracts. Because data limitations did not allow for identification of individual prospects in other USGS provinces, plays were treated in the composite fashion described above.

Geologic Risk Factors

The play appraisal form required the assessment of separate risk factors for source, timing, migration, facies, and minimum accumulation. Each category was assigned a single value between 0 (total certainty that the attribute is absent) and 1 (total certainty that the attribute is present). The USGS treated these values as statistically independent and multiplied them together.

In the case of the hydrocarbon source attribute(s), the USGS reportedly considered organic richness, kerogen type, and thermal maturity interactively in arriving at some value between 0 and 1. However, if oil or gas in commercial amounts was present in the play, then the play attribute probability was automatically assigned a value of 1. Similar determinations apply to the assignment of the attribute probabilities for timing (T) and migration (M), with a value of 1 being assigned if accumulation had occurred. As a consequence, for all plays that had identified in-place petroleum accumulations, regardless of the amount or quality, attribute probabilities for S, T and M were assigned a value of 1. Thus, the USGS ignored geochemical data in estimating resource volumes when petroleum had already been found in the area being assessed. The committee recommends that the USGS determine the potential impact on the assessment of omitting geochemical data in assigning risk to source, timing, and migration for plays with in-place petroleum accumulations.

Most interior onshore basins have established volumes of oil and gas. Therefore, the general geochemical premise that the USGS adopted is that there is oil and/or gas present and hence there are source rocks. This premise generally permeated the entire assessment process.

Although presence or absence of oil and gas source rocks is correctly used to define whether an area is *prospective* for petroleum, the presence of petroleum source rocks is not a stand-alone indication that oil and gas accumulation has actually occurred in contiguous reservoir facies within the play. One must further integrate into the assessment process the interrelationships between the potential source facies and the contained reservoir facies. Most important, one must consider the degree of geothermal diagenesis that the sedimentary source and reservoir sections have experienced throughout geologic time. The USGS did not consider these basics consistently.

If a specific play is located in thermally immature sediments, then any contiguous reservoirs within that play can be prospective only for either indigenously sourced biogenic gas or for migrated oil and/or gas. The chance that indigenously sourced oil is present in thermally immature reservoirs is zero. Clearly, the chances for any oil and gas accumulation in the play will depend on migration of petroleum into the play, which in turn will depend on source rocks being located outside the play. In a like manner, if the play in question is associated with a sedimentary section that is thermally mature, then all associated contiguous reservoirs within the play may be prospective for both indigenously sourced and migrated oil and gas. Finally, for very mature strata, prospective target reservoirs should be candidates for indigenously sourced, thermally generated gas accumulation. Thus, play risk and the amount of oil and gas in place are functions of the play's thermal maturity regime. Consequently, the play's thermal maturity—immature, optimally mature, or very mature—should be clearly defined.

The committee concludes that formal consideration of thermal maturity and its inclusion in the assessment forms may substantially alter future projections of undiscovered oil and gas. The USGS should test this proposition.

Equally important as maturity consideration in play analysis is the proper definition of the organic facies characteristics associated with each play. It is insufficient to conclude that generally oil-prone organic matter and generally gas-prone organic matter will always yield the expected product. The committee recommends that in future assessments, analysts use geochemical data wherever possible to evaluate conceptual and emerging plays and established plays in which source and maturity considerations affect prospect risk. In evaluating source and maturity risks, the USGS should use its own extensive geochemical research and other recently published research.

Sizes of Undiscovered Fields in a Play

The appraisal form required that analysts input separate size distributions for undiscovered oil and undiscovered gas fields. As explained in Chapter 2, analysts created these size distributions by fitting the sizes of already discovered fields in a play to a truncated shifted Pareto (TSP) distribution. According to the USGS, analysts chose a TSP distribution in place of the more commonly used truncated, three-parameter lognormal distribution because the TSP distribution is mathematically simpler and generally yields results similar to the lognormal distribution.

To assess how the sizes of fields discovered change through time, analysts created three chronological TSP distributions for each play: one for the first third of fields discovered, one for the middle third, and one for the last third. In one Permian Basin gas play, for example, the first TSP distribution was based on the sizes of all fields discovered before 1968, the second covered fields discovered between 1968 and 1973, and the third included fields discovered between 1973 and 1983. Presumably, the parameters of the TSP distribution evolve with time. Thus, USGS analysts used the chronological TSP distributions created from known fields to extrapolate parameters for a TSP distribution for the undiscovered fields in the play (Houghton, 1988). The group responsible for evaluating the play considered the extrapolated parameters, but did not necessarily use them to define the final undiscovered field-size distribution for the appraisal form. The final undiscovered field-size distribution was based on the analysts' interpretive evaluation of the three components of the discovery record.

The committee recommends that the USGS investigate whether the manner in which the TSP distributions fit the discovery record is sensitive to the fashion in which the data are partitioned. Choice of division into two or four in place of three time segments may result in very different fitting patterns. This, in turn, would influence the subsequent judgments of geologists in their assessment of a shape for the size distribution of undiscovered deposits.

Analysts used a point estimate of the size of the largest field to calibrate the fit. This largest size depends on the *number* of undiscovered fields (i.e., it is an "order statistic"). Thus, the fitting procedure introduced two forms of dependence: dependence of the undiscovered size distribution's shape on the assessment of the number of undiscovered fields and a consequent probabilistic dependence between the uncertain quantities "undiscovered field size" and "number of undiscovered fields."

The committee judges that it is a logical contradiction to treat the uncertain quantity "undiscovered field size" as independent of "number of undiscovered fields" in subsequent calculations. Furthermore, the notion that number and size should be treated as statistically independent is contrary to reasonable models of the discovery history of a play. The undiscovered field-size distribution would be assumed the same whether there were 30 or 300 undiscovered fields in the play.

Number of Undiscovered Fields in a Play

The USGS's procedure for assigning the distribution for the number of undiscovered fields in a play is poorly described and seems unrelated to the assigned field-size distributions. While the USGS invested considerable effort into structuring the discovery history as it relates to the evolution of the field-size distribution, it apparently made no parallel effort to couple this with a parametric structuring of a corresponding distribution for the number of undiscovered fields.

For each play, the mean and variance of the assigned distribution for the total number of hydrocarbon accumulations was somehow partitioned into a separate mean and variance for the number of oil fields and for the number of gas fields. The one piece of information used for this purpose was an assigned oil/gas probability, but the formula for partitioning variance is not obvious nor is it documented. Presumably, the decision about whether an accumulation is gas or oil resulted from a binomial trial using the assigned probabilities. The committee judges that in future assessments, a better approach would be to use separately specified distributions for the number of undiscovered oil fields and the number of undiscovered gas fields in a play.

Aggregation Within a Play

The USGS characterized each play's aggregate resources with a mean and variance. Analysts obtained the mean and variance by considering the aggregate as a sum of a random number of randomly sized fields, treating the number and size as statistically independent quantities. For this calculation, evaluated above, the only inputs needed were the means and variances of the number and size distributions for the play. Similarly, only the play aggregate mean and variance were required for the higher levels of aggregation.

USGS analysts could add to the play aggregate additional resources from small fields (those containing less than 1 Mmboe). It is not clear how the USGS propagated small field uncertainty to higher aggregation levels. Overall, small fields contributed about 27 percent to the USGS's total national resource estimate.

Analysts made judgments about the chance of occurrence of the geologic conditions necessary to have formed at least one remaining significant undiscovered accumulation in the play. Analysts then proceeded quantitatively to assess accumulation sizes and numbers as probability distributions. As described previously, a truncated shifted Pareto distribution was fitted to the sizes of the known oil and gas fields in each play to guide the analysis of field-size distributions. Geologists performed the resource calculation by means of FASPF (Fast Appraisal System for Petroleum Field Size) using an analytical method based on fitting lognormal distributions with the specified means and variances, rather than using a Monte Carlo simulation (Crovelli *et al.*, 1988; U.S. Department of the Interior, 1989).

One problem with the USGS procedure for aggregating resources in a play was that it distilled evaluations of several analysts into one consensus result. Consensus was used to obtain both geologic risk factors and probability distributions showing the range of possible resource volumes thought to exist in the play. The USGS provided no detailed description of the protocol employed to arrive at consensus values. In addition, it reported only the consensus risk factors and resource distribution, masking the varied assessments of the participating geologists. The result is a compression of uncertainty: a failure to reflect the diversity of opinion between individual geologists in the published estimates. Consensus distributions usually concentrate on a range of values smaller than that spanned by the union of individual opinions. The history of oil and gas forecasting teaches us to be wary of point estimates equipped with narrow credible probability intervals.

The committee judges that, especially in frontier areas, the range of uncertainties in resource numbers should reflect the uncertainties exhibited in the differing opinions of geologists. We recommend that, at the very least, the range of risk factors assessed by individual geologists before a consensus is reached be available for assessment users to inspect.

Another problem with the method for aggregating resources in a play was that it treated the attributes that lead to petroleum accumulation—source, timing, migration, and reservoir—as statistically independent. Analysts used subjective judgment to determine the probabilities that these attributes were present. Assuming the attributes were independent, analysts multiplied the probabilities together to yield the marginal probability that a play contains a certain minimum petroleum volume. In most plays, however, source, timing, and migration are related geologically and therefore may be dependent. Treating the factors as independent may have yielded a lower marginal probability than if they had been treated as dependent. The precise impact of modifying this independence is poorly understood because no alternatives have been tested. However, in Alaska's North Slope province, which includes several multi-billion-barrel plays, treating accumulation attributes as independent could have significantly influenced the projection of oil and gas volumes.

The committee recommends that the USGS test the sensitivity of its undiscovered oil and gas projections to the assumption that petroleum source, timing, and migration are statistically independent.

Aggregation of Plays Within a Province

In this assessment, the USGS treated undiscovered resources in plays within a province as perfectly correlated. For example, if one play's resources exceeded the mean of its distribution by 1.5 standard deviation units, then the USGS assumed that every play in the province exceeds the mean of its distribution by exactly 1.5 standard deviation units. Under the assumption of perfect correlation, the standard deviation of the uncertainty distribution for a province becomes equal to the sum of the standard deviations of the uncertainty distributions of the component plays. Thus, in the USGS method, only the province mean and standard deviation were needed for the higher aggregation levels.

The committee is concerned that the method the USGS used to aggregate resources within provinces is inconsistent with the method the MMS used. While the USGS method assumed complete conditional dependence among plays, the MMS method (evaluated later in this chapter) assumed complete (conditional) independence. The committee recommends that in future assessments, the USGS and MMS standardize their procedures for aggregating resources.

Regional and National Aggregates

USGS assessors estimated the standard deviation of the uncertainty distribution for regional aggregates as the exact mid-value between an assumption of perfect correlation among component provinces and an assumption of zero correlation. Thus, the degree of correlation among provinces was not region-specific. Similarly, the aggregate mean was always assumed to be the aggregate of the component means; it was not tied to distributional or correlation assumptions.

The regional aggregate is the first aggregation level for which the USGS reported uncertainty estimates in terms of percentiles. Therefore, at this level, it was necessary to describe the shape of the uncertainty distribution and not just its mean and standard deviation. This was done by assuming a lognormal distribution with matching mean and standard deviation and reporting the percentiles of the fitted lognormal distribution.

The USGS's approach to aggregation thus rested on carrying forward standard deviations for successive aggregation levels via correlation assumptions, and on a final imposition of a lognormal distribution at the regional level. Presumably, the USGS adopted this approach to avoid conducting a simulation based on the actual size and number distributions specified at the play level. Yet, the MMS used a simulation approach that does not appear especially burdensome. Because uncertainty percentiles are not reported below the regional level, there is no need to develop distributions for aggregates below this level. Consequently a simulation approach would require that individual play aggregates be simulated only once for each simulated value of the regional aggregate.

The committee recommends that the USGS adopt a simulation approach to aggregation at the regional level. The national aggregate uncertainty distribution can easily be obtained from the regional uncertainty distributions by a simulation, because the number of regions is not large and independence between regions is a plausible hypothesis. The advantage of simulation is that all the uncertainty information contained in the original play appraisals can be carried forward. Therefore, assessors can avoid adoption of a particular parametric form for the regional uncertainty distribution. Before the USGS can implement a simulation approach, however, it must give further thought to the degree of dependence among plays within provinces and among provinces within regions. The MMS introduced dependence by using a hierarchical risking structure, but this may not be appropriate where risks are essentially zero.

Subjective Probability Assessment and Training

The use of personal probabilities in the assessment of quantities of undiscovered oil and gas is well established. Two notable examples of government agency projections of undiscovered petroleum that use subjective probability are described in "Geological Estimates of Undiscovered Recoverable Oil and Gas Resources in the United States" (Circular 725), published by the USGS in 1975, and "Oil and Gas Resources of Canada, 1976," published by Energy, Mines and Resources, Canada.

The recent USGS assessment method differed from that described in Circular 725 in two important ways: it adopted the petroleum play as the basic unit for analysis and it used key features of discovery process modeling as a guide to formulating subjective assessments. As discussed earlier in this chapter, each geologist subjectively appraised the size distribution of a play's undiscovered fields by choosing the parameters of a truncated, shifted Pareto distribution. The idea that field size declines on average as exploration progresses was incorporated by displaying three, chronological TSP distributions: one representing fields discovered in the first third of the play's history, another representing the middle third of the discoveries, and another representing the most recently discovered fields.

The most notable feature of this assessment protocol is that it employs properties of an objective first principles model of discovery as an aid in making subjective assessments *without* exploiting the predictive capabilities of such a model. Each assessor must use his or her judgment to appraise a descriptively complex, compound event and create distributions for the sizes and numbers of undiscovered fields. In contrast, an objective first principles model of discovery is designed to generate this distribution as a function of the discovery history *without* subjective intervention.

Among the issues raised by the USGS's assessment approach are:

1. Is the TSP distribution, which assessors use as a decision aid, adequate to represent the observed discovery histories for the wide spectrum of petroleum plays that the assessment covers?
2. How well trained are participating geologists in subjective probability assessment?
3. Do the results adequately represent uncertainties about field-size distributions and numbers of undiscovered fields?

Houghton studied the quality of the fit of a TSP distribution to the discovery history of a very mature play, the Minnelusa play, in the Powder River Basin and compared this fit with some alternatives (Houghton, 1988). Houghton's study appears to be the only attempt, prior to the national assessment, to validate the model on which the USGS assessment procedure is based. Given the geological diversity of the approximately 250 plays included in the national assessment, a much more aggressive effort to validate the procedure prior to its adoption was warranted. The committee questions whether the USGS's assessment procedure is robust, in the sense of providing a good fit to widely varying discovery histories, and precise, in terms of providing a good fit between predictions of future discovery patterns and existing discovery patterns.

Discussions with assessors suggest that the USGS devoted minimal effort to the organized training of participating geologists in subjective probability. In response to a question about the level of training, a senior manager replied: "Most of the geologists have probably attended a short course in subjective probability assessment." Because of the lack of training, it is not surprising that in interviews with participating geologists, we found many examples of failure to understand the precise meanings of fundamental probability concepts. In particular, widespread misunderstanding of the meanings of .05 and .95 fractiles, maximum possible and minimum possible values, risk, and conditional versus unconditional probabilities appeared during our interviews.

The committee found that these shortcomings of the assessment procedure are symptomatic of the USGS's failure to maintain support for research and training over the long time periods necessary to carry out a credible oil and gas assessment. Because of the absence of a permanent resource assessment group, the assessment lacked continuity in structure, a clearly defined assessment methodology that was unambiguously understood by all members of the assessment team, and enough statistical support to carry out data management, modelling, and predictive validation.

Case Study: Alaska

Alaska is perhaps the most important U.S. region for future onshore petroleum discovery. The 1989 assessment concluded that onshore Alaska and state offshore Alaskan waters house 26.9 percent of the nation's technically recoverable undiscovered petroleum. Because of this region's significance, the accurate appraisal of its resources is critical for determining the nation's future energy strategy. Consequently, the committee analyzed the assessment more closely in Alaska than in other regions. This section presents the committee's findings related to the USGS's appraisal of Alaska's potential resources.

Overall, the committee concluded that the quality of the USGS assessment in Alaska was good compared to some other regions. In large part, the credibility of the USGS's work there was due to the staff that covered the region. The assessment team that evaluated Alaska was large compared to the teams for some other regions, and many staff members had prior resource assessment experience. Nevertheless, the committee identified some areas of the assessment process that could be improved.

Data Base

The USGS's geological data base is more complete in Alaska than in many other regions. In particular, the region covering Alaska's National Petroleum

Reserve (NPRA) has perhaps the most complete geochemical data base available. In part, this is because in Alaska, unlike in other regions where much of the well data are from industry and are proprietary, the USGS has access to data from almost all wells (except for a small number of North Slope wells for which information was withheld because of a state confidentiality period). Also, in Alaska, all the engineering and production data dating back to the beginning of production in 1957 fall under the USGS's purview. And, since the last century, the USGS has been collecting comprehensive geological data for the region.

In contrast to the geological data base, the USGS's seismic data base for Alaska is less complete. For this assessment, seismic data coverage was limited to the Arctic National Wildlife Refuge (ANWR), the NPRA, and a single, pre-common-depth-point (CDP) seismic line through Prudhoe Bay. The committee is concerned that the incomplete seismic data base, especially for the North Slope, creates an imbalance in play analysis and resource assessment in an area with one of the highest resource potentials in the onshore United States.

One way the USGS might augment its seismic data base is by increased cooperation with state government agencies and the petroleum industry. For example, it is possible that the USGS could persuade industry to allow province geologists to view proprietary seismic lines in key areas.

Play Methodology

As discussed earlier, two of the committee's concerns about the assessment relate to how the USGS defined plays and how it accounted for conceptual plays. Improper play definition can result in play mixing: the combination of dissimilar depositional systems into one category, which prejudices statistical computations and may skew the assessment toward an underestimation of resource volumes. The failure to consider conceptual plays—those without discovered petroleum but that geologic information indicates may exist—can also lead to an inordinately low resource estimate.

The committee evaluated in detail the 12 plays the USGS identified on the North Slope. From this review, the committee concluded that North Slope plays were, for the most part, well defined. Generally, prospects and discovered pools grouped into plays shared the requisite common history of petroleum generation, migration, reservoir development, and trapping conditions (Podruski *et al.*, 1988). Nevertheless, some mixing of plays may have resulted from the inclusion of thick, stratigraphic sequences with different lithologies and origins in the same play. For example, the Barrow Arch Play includes the entire Ellesmerian sequence, which consists of rocks ranging in age from Mississippian to Lower Cretaceous, deposited in both marine and non-marine environments and variously composed of sandstones, conglomerates, shales, and carbonates.

In contrast to the fairly consistent definition of North Slope plays, from the committee's evaluation it appears that the USGS's consideration of conceptual plays was sporadic. The USGS defined conceptual plays in some parts of Alaska, but not in others. For example, in the North Slope an "economical top of Lisburne" became a conceptual play. However, in the Cook Inlet province, no conceptual plays were included, though members of the Alaska panel were aware of conceptual plays identified and being pursued by industry.

In future assessments, the committee recommends that the USGS attempt to identify conceptual plays throughout Alaska. Because of the major potential for uncertainties in evaluating conceptual plays, however, it is essential that the volumetric contributions of conceptual and established plays be clearly separated. In regions where the USGS included conceptual plays in its resource estimates, the committee found inadequate documentation of how much of the estimates came from conceptual plays and how much from established plays.

Technically Recoverable Resources Computation

For North Slope plays, the USGS produced its estimate of the technically recoverable resource volume by applying a "recovery factor" to its estimate of the total petroleum in place. This recovery factor was uniform across all plays (32.3 percent for oil). Because the recovery factor percentage is a direct multiplier in converting oil-in-place to reserves, small variations in the percentage applied can change North Slope totals by billions of barrels. It would therefore be more appropriate to evaluate each play, or even sub-play, according to its particular characteristics. Across a region as large as the North Slope, these characteristics will vary considerably with lithology, depth of burial, tectonic setting, and applied drilling technology. For future assessments, the vast amount of production history and data now available from the Endicott, Lisburne, Prudhoe Bay, and Kuparuk fields and the greater density of infill drilling and horizontal drilling should improve the determination of recovery factors.

Economic Assumptions: Minimum Economic Field Size

To determine economically recoverable resource volumes, the USGS applied a single minimum economic field size (MEFS) of 384 Mmboe on the

North Slope for all plays in all geographic locations. The rationale behind this assumption was that 384 Mmboe is the mid-point of the area's field-size distribution, which ranged from 256 to 512 Mmboe. Assessors determined that fields at the bottom end of this range—those that contained 256 Mmboe—were not commercially developable.

After its review, the committee concluded that the 384 Mmboe cutoff may be too severe. For example, the MEFS of accumulations close to the pipeline should be substantially smaller than the MEFS of farther-removed accumulations because of the lower transportation costs. In addition, the committee concluded that the use of a single MEFS for the entire North Slope is difficult to justify. The use of a range of MEFS values more tailored to specific areas is possible without extensive additional effort. For example, the MMS, in its evaluation of federal offshore waters in the Beaufort and Chukchi Seas, used MEFS values ranging from 208 to 278 Mmboe for "satellite" fields and from 517 to 810 Mmboe for "stand-alone" fields. For future assessments, the committee recommends that the USGS study and refine the MEFS values it chooses for the North Slope. Cooperative studies between the USGS and the MMS could result in more realistic and consistent economic screening.

Uncertainty

In Alaska more than in perhaps any other region, the USGS should emphasize the uncertainty in its resource estimates. Though Alaska contains a large fraction of the nation's petroleum reserves, compared to other regions its exploration and production history is immature. Even in the generally accepted "mature" areas of Cook Inlet and Prudhoe Bay, production began relatively recently, in 1957 and 1968, respectively. The comparatively unexplored nature of many areas in Alaska magnifies the already substantial uncertainties that surround predicted resource volumes. The committee recommends that the USGS assume the responsibility of communicating this uncertainty to assessment users, making certain users do not limit their focus to mean, single number estimates, but instead consider the entire range of possibilities.

DETAILED EVALUATION OF MMS ASSESSMENT METHODS

Until the DOI established the MMS in 1982, the USGS had the responsibility to assess oil and gas volumes in federal offshore territory. Before the MMS

BOX 3.2 The MMS'S Legislative Mandate

The MMS is a relatively new player among the federal agencies. It was established in 1982 to consolidate the administration of mineral resources in federal offshore territory. Prior to the establishment of the MMS, management of offshore resources was divided among several government branches, including the USGS, the Bureau of Land Management, and the DOI's Office of Outer Continental Shelf (OCS) Program Coordination.

Unlike the USGS, the MMS has a legislative mandate to submit biennial reports to Congress that assess the undiscovered economically recoverable resources of the OCS. This mandate is specified in 1978 and 1985 amendments to the OCS Lands Act of 1953.

Assessing mineral resources is only one part of the MMS's responsibilities. Its primary task is to manage the federal government's leasing of OCS mineral resources, as authorized in the Lands Act. The MMS approves leases of offshore acres for petroleum exploration and collects royalties from companies as payment for offshore leases.

was formed, management of Outer Continental Shelf (OCS) resources was split between several divisions within the DOI, including the USGS and the Bureau of Land Management. But as interest in producing offshore petroleum to reduce reliance on imports grew, the government decided that the authority over the OCS should be consolidated in one agency: the MMS. Thus, the MMS is charged with assessing petroleum resource volumes on the OCS (see Box 3.2). It is required to report these volumes to Congress every two years. This section evaluates the quality of the MMS's work for the 1989 assessment.

Organization and Staff

To fulfill its legislated responsibilities of performing biennial resource assessments, the MMS has subdivided the OCS into four regions: Alaska, Pacific, Gulf Coast, and Atlantic. A semi-autonomous office oversees each region. Regional offices are located in Anchorage, Alaska, Los Angeles, California, New Orleans, Louisiana, and Herndon, Virginia. In addition, the MMS has a headquarters Office of Resource Evaluation in Herndon.

Because the MMS has a legal obligation to perform offshore resource assessments biennially, and because economic resource evaluation is a continuing process for federal offshore lease sales to petroleum companies, the MMS maintains a large, exceptionally experienced professional staff in all four regional offices. (Altogether, more than 400 MMS scientists contributed to the 1989 assessment.) Experience levels are sufficient to provide sound knowledge of the various geological provinces within the regions. For example, the resource group in the Gulf of Mexico OCS region has a permanent professional staff of seven, most of whom have advanced degrees in geology or geophysics. Engineering, statistical, and mathematical specialists normally are assigned on a regional basis and are available to the province or basin groups. Many of the MMS professionals have had industry experience. Because of the obligation for biennial assessments, assessment in most areas is ongoing. Thus, there were no apparent time constraints for the 1989 assessment. Most of the work was accomplished with proper attention to detail and thorough analysis.

The experience level of the MMS staff increases the credibility of its resource assessments. The committee is concerned, however, that funding cutbacks caused by moratoria on lease sales will force the MMS to reduce personnel levels, which could have a negative effect on its ability to conduct future assessments.

For the most part, staff training and background for the 1989 assessment were adequate to enable the staff to interpret data competently and to translate their interpretations into resource assessments with credible results. Nevertheless, there is evidence that MMS geologists shared difficulties experienced by USGS geologists in understanding the precise meaning of some important probability concepts. For example, one form (Form B) that the MMS used in Alaska asked for an appraisal of "minimum" and "maximum" values for several uncertain quantities to be assessed by geologists. At a meeting with our committee, some geologists said they interpreted the minimum value as the .05 fractile, while others thought it was a value such that there is zero probability that the uncertain quantity will fall below it. The maximum possible value was also interpreted inconsistently. Confusion about the operational meaning of elementary probability concepts may be limited to the group of geologists we met. Nevertheless, the committee advises that in future assessments, the MMS should ensure that all participating geologists have a clear understanding of probability concepts that they must employ. Otherwise, the credibility of the assessment numbers may be questioned.

This small piece of evidence suggests that some geologists may have experienced difficulty in assessing probabilities for descriptively complex, compound events. Their assessments may not have been well calibrated with (i.e., may not have corresponded with) actual relative frequencies of such events. The MMS could conduct controlled experiments to measure how well geologists' assessments calibrate with actual events.

The committee has a related concern about the vagueness of certain crucial terminology used in the assessment forms. For example, Form B asked for the probability of the presence of a "suitable reservoir." The measurable attributes that define "suitable" were not defined in writing. In practice, geologists assessing this uncertain quantity may have shared a precise understanding of these attributes. However, without documentation, it is possible that "suitable" was interpreted differently by different geologists. If so, probabilities assigned by different geologists to "suitable reservoir" were not comparable. The committee recommends that the MMS check that all the terms on its assessment forms are well defined.

Lease sale economic evaluations constitute a major part of MMS responsibilities. Although the MMS conducts resource assessments independently from lease sale evaluations, it is apparent that lease sale processes directly affect assessment procedures. For the 1989 assessment, priorities within the MMS limited detailed economic analysis of prospects mostly to tracts that had received bids. Greater MMS emphasis on resource-base assessment processes could provide more definitive knowledge of the resource base for long-term exploration under different political, technological, and economic scenarios. This, in turn, could provide better support for enlightened policy formulation. Consequently, the committee recommends that the MMS map and evaluate in detail *all* prospects, not just those associated with lease sales.

In summary, the committee generally views the MMS staff scientists as qualified and well organized. They are competent and capable of applying currently accepted methods to their analyses and interpretations. Nevertheless, the committee is concerned about the staff's lack of a common understanding of probabilistic terminology and concepts as applied in the assessment. The quasi-autonomous nature of management of the MMS regions does not lend itself to coordination between offices and hence can lead to differences in definitions, procedures, and risking criteria. The committee is also concerned that the MMS lease sales exert too much influence on deciding how thoroughly to evaluate prospects.

Data Base

Since the MMS has the responsibility to regulate exploration and development activities on the OCS, it has gathered an extensive data base of geological, geophysical, engineering, and production information. As part of the permit and lease terms, the MMS acquires proprietary geological and geophysical information collected by industry.

The MMS has acquired over 1 million line miles of common-depth-point (CDP) seismic data: the Alaska OCS is represented by an estimated 332,500 miles, the Pacific by 99,000 miles, the Georges Bank area of the Atlantic by 62,000 miles, and the Gulf of Mexico and the Atlantic OCS south of Georges Bank by the remainder. The MMS's seismic acquisitions consist of lines that have been processed; the MMS does not have reprocessing capabilities. The seismic data base includes three categories of mapping density: (1) reconnaissance mapping with wide seismic grids of 2 to 20 miles, (2) basin-wide mapping, and (3) lease-sale-related mapping with 0.5- to 1-mile grids to detail specific prospects. The highest density of seismic grid patterns is concentrated on and around acreage blocks associated with lease sales. In other areas where there has been little or no leasing or active exploration, such as the Northern California planning area, part of the eastern Gulf of Mexico area, and some parts of the Atlantic OCS, seismic data are sparse to non-existent.

In addition to the seismic data, the MMS has access to data from the more than 25,000 wells drilled in federal waters. Information from these wells includes geophysical logs, cores, cuttings, formation and production test records, well histories, and limited geochemical data. Included are 67 exploratory wells and 4 Continental Offshore Stratigraphic Test (COST) wells off Alaska; some 925 wells in the Pacific OCS; 49 exploratory wells and five COST wells in the Atlantic OCS; and thousands of exploratory and development wells, including two COST wells, in the Gulf of Mexico.

Though industry data from wells and seismic tests provided the MMS with a fairly solid foundation for its 1989 assessment, there were gaps in the geochemical data. For example, though the MMS used COST wells to evaluate the oil and gas source-rock characteristics of Alaskan offshore basins, these evaluations were mostly "one-well" basin assessments and consequently had limited areal coverage. Industry has, in fact, drilled key exploratory wells in Alaskan offshore basin areas. However, the MMS does not presently mandate that industry conduct a full, COST-type geochemical well evaluation for these permitted wells for later use in ongoing MMS assessments. Similarly, in the Baltimore Canyon in the Atlantic, the MMS relied on information from two COST wells to assess the potential for petroleum resources, although industry had drilled a multitude of exploratory wells. The MMS did not mandate that the industry wells, which could have provided valuable data, be analyzed as COST wells.

The committee recommends that for wells selected for drilling in offshore areas of sparse control, a full compliment of samples should be collected and subjected to a full COST well analytical program. This would allow the MMS to update continuously its evaluations for those areas and would expand the data base for future assessments. One good example of where the MMS has already made full use of industry well data is in the Georges Bank area. There, two industry-sponsored COST wells were drilled at geologically distinct locations: one penetrated a clastic sequence near the shore and the other penetrated the offshore carbonate section. A number of industry wells drilled after the initial COST wells were subjected to a similar geoscientific evaluation. As a result, the MMS employed the full data set available from these wells in its assessment of this environmentally sensitive area.

The MMS has access to an onshore data base containing published material from federal and state agencies, industry, and academic sources that might also benefit the USGS. Engineering and production information is available from the inception of production both onshore and offshore from federal and state agencies. In cases where little or no exploration activity has taken place offshore, analogues from geologically similar petroleum provinces onshore and in other parts of the world are incorporated into the assessment procedures.

The committee recommends that in future assessments, complete cooperation between the MMS and the USGS, as sometimes occurred for the 1989 assessment, be established and maintained. Direct involvement of other federal agencies, such as the BLM, and pertinent state agencies also will enhance results of future studies. For example, the California State Lands Commission retains control over confidential data in the 3-mile zone of California state waters. For the 1989 assessment, this caused a critical gap between onshore and offshore projections. The USGS and MMS should undertake a joint, concerted effort to obtain the data. Similar gaps exist in other areas and should be filled.

Play Analysis

The MMS identified 74 plays for the 1989 assessment: 34 in Alaska, 3 in the Pacific, 24 in the Atlantic, and 10 in the Gulf of Mexico. Assessors categorized plays as either as "identified" (predominantly structural plays identified from seismic interpretations) or "hypothetical" (predominantly stratigraphic plays postulated because of information about onshore geology or data from similar provinces or basins). In frontier areas, extensive use was made of published literature for an understanding of the geologic setting and play potential.

For assessing offshore resources, delineating plays according to the precise definition of a play is difficult because offshore exploration is relatively new and therefore data are limited. Strictly, prospects should be grouped into plays based on the geologic similarity of their source-reservoir-trap controls. The MMS did not always follow precise play definition, sometimes delineating plays by subdividing a region stratigraphically, such as in the Pacific Coast region. Despite gaps in data, the committee advises that the MMS adhere more strictly to the definition of a play in grouping prospects for future assessments.

Play Definition: Potential for Play Mixing

For the 1989 assessment, the MMS's definition and application of plays varied within each region and among regions. In the Atlantic region, the plays appear to have been defined rigorously. Assessors grouped prospects into plays based on geologic setting, structure, and stratigraphy. Similarly, Alaska plays appear to have been well-founded, although we detected some mixing of plays. For example, in the Chukchi Sea, the Western Ellesmerian Play included both carbonate and clastic lithologies, ranging in age from Mississippian to Lower Cretaceous. (Given the area's early stage of exploration and the consequent lack of data and knowledge, this broad inclusion of lithologies and ages may not be have been unreasonable.) In the Pacific Coast region, a three-fold, stratigraphic subdivision was used as a basis for play delineations: pre-Monterey, Monterey, and post-Monterey. The process was consistent within the region but did not conform with play definition as used elsewhere in the assessment. In addition, the gross three-fold subdivision of the entire stratigraphic section between the Canadian and Mexican borders created the potential for mixing of plays. In the northwestern Gulf of Mexico, plays were defined broadly, based on the ages of productive reservoirs or potentially productive trends. Thus, Gulf of Mexico plays included a wide range of geologic and tectonic settings, lithologies, stratigraphies, depositional systems, and structural characteristics, making the mixing of plays unavoidable.

The effect of play mixing in parts of Alaska, in the Pacific Coast region, and in the Gulf of Mexico cannot be determined without specific tests to examine the impact of mixing in each region. However, in the cases we have examined, play mixing resulted in overly conservative resource estimates by obscuring gaps in field-size distributions. (For a further discussion of the possible effects of play mixing on resource estimates, see the discussion of the USGS's application of play analysis, earlier in Chapter 3.) For future assessments, the committee recommends that the MMS develop geological and statistical procedures to insure that play mixing does not significantly alter resource estimates. The MMS should test results obtained by disaggregating plays, particularly those containing diverse lithologies, sedimentary origins, or facies. To create better play definitions and prevent play mixing in California, the MMS should evaluate all potential source beds and migration paths instead of focusing almost exclusively on the Monterey formation.

Conceptual Plays

Because of the extensive seismic data base available for OCS evaluations, the MMS's play definition procedure for the 1989 assessment consisted largely of structural prospect identification. Potential productive zones were aggregated within each prospect, prospects were aggregated into plays, plays into provinces (or basins), and provinces into regions. This approach depended entirely upon the density and quality of the seismic grid and the interpretive skills of the geoscientists working in the individual areas. In the areas where the seismic grid was too widely spaced to identify and map structural closure, estimates of numbers of prospects were generated by analog or statistical procedures. These were defined as "unidentified" prospects or as "conceptual" prospects or plays. However, the committee could not identify any explicit procedure in the assessment whereby the MMS evaluated coverage and density of seismic surveying, drilling, and other sources of data for prospect identification and estimated the potential for unidentified resources on the basis of the volume and location of rock not included in the survey coverage. The committee judges that the manner in which the MMS defined conceptual plays did not result in the proper identification and categorization of such plays, and thus missed their potential contribution to the overall resource endowment.

In the Pacific Coast region, using all available subsurface well data and extensive seismic reflection data, MMS personnel did attempt to define, map, and estimate undiscovered resources for "postulated" or "unidentified" plays. These included stratigraphic traps (e.g., pinch-outs or truncations), complex structures (e.g., thrust faults), diagenetic alterations, and statistical projections of unmapped structural closures in areas with widely spaced seismic grids. Nevertheless, it does not appear that MMS gave appropriate weight to conceptual plays even here. The committee report reprinted in Appendix A noted that:

> in the northern California Eel River Basin, the entire "postulated" resource was allocated to only an estimated four "unidentified" structural prospects as compared to 92 "identified" structural prospects. In the entire basin the "unidentified" category yielded a risked total resource of only five million barrels of oil as compared to a total of 159 million barrels of oil for the "identified" category. This process does not include possible stratigraphic or other types of conceptual traps known to occur in other California basins.

In the same report, the committee concluded that in California, the "focus upon the Monterey Formation as the dominant source (and reservoir) may have the effect of downplaying other potential hydrocarbon source beds and thus tend to understate their contribution to postulated or unidentified plays in resource assessments."

In Alaska, allowance for the existence of conceptual plays was also a part of the evaluation process, but this allowance was generally applied at the province level and was not play-specific, nor was a specified portion of the resource estimate allocated to conceptual prospects/plays in most provinces. Thus, the contribution of conceptual plays to the resource base may not have been identified or may have been missing entirely. The economics of developing and producing high-risk, difficult-to-identify conceptual plays, such as stratigraphic traps, tend to remove them from the economically recoverable resource estimates in Alaska. This is the dual effect of extremely high-cost development in the Alaska OCS, combined with the risk factors that cause such plays to fail to meet MEFS and/or minimum basin reserves (MBR) criteria. In the Alaska OCS, stratigraphic traps are often elongated and have relatively thin reservoirs. They therefore require more platforms with fewer wells, thus defeating economies of scale that accrue to compact reservoirs with thick pay sections. It is, however, the responsibility of MMS to identify and assess such potential plays in the recoverable resource endowment.

Aside from the MMS's limited treatment of conceptual *plays*, the MMS's treatment of unidentified *prospects* in *identified* plays also had strong limitations. In Alaska, for example, only identified prospects were considered to comprise the inventory of prospects characterizing the resource base. The collection of unidentified prospects that may be present was ignored. In other offshore areas, a point estimate rather than the possible range of the number of unidentified prospects was adopted as a surrogate for judgmental assessment of the probability distribution of this very uncertain quantity. This created a less severe but still important compression of uncertainty.

It is not surprising that MMS assessors in Alaska focused on identifiable structural prospects and their aggregation into plays, given the total lack of economic production in this region. Without some evidence that *some* recoverable resources lie outside of identified structures, it may be difficult for them to justify postulating such plays. Some inclusion of conceptual plays has occurred, however, as in the Navarin Basin "Flank Sands" play.

Nevertheless, in the committee's view, the small number of conceptual plays the MMS identified in the Alaskan offshore was inconsistent with this region's development stage and with the availability of exploration data. The number of conceptual plays generally should be at a *maximum* early in the history of a basin, and decline with time as more of the basin's rocks are tested with both more seismic data and drilling. During the early years of basin exploration, the number of conceptual plays should fluctuate significantly with time as the continual gathering of new evidence reduces the potential for some types of plays and/or increases the potential for others. For example, an initial set of stratigraphic discoveries might imply that much of the untested acreage could contain other stratigraphically trapped resources—though assessors might consider this acreage to be part of the already identified plays and thus would not add new conceptual plays to the inventory. Conversely, new findings about basin temperature histories might challenge previously postulated oil plays, reducing the number of conceptual plays. As the region matures and the data base increases, many conceptual plays will be tested and will either prove dry or be moved to the identified category.

For the 1989 assessment, the MMS included conceptual plays in the Atlantic OCS evaluation. In the Gulf of Mexico, however, although there is widespread seismic coverage of relatively unexplored areas showing the existence of allochthonous salt sheets and tongues (Morton and Nummedal, 1989), and possible traps related to these, we found no mention of conceptual plays related to such features in the assessment, nor were other conceptual plays postulated.

In summary, the committee concluded that the MMS should have identified many more conceptual plays, especially in Alaska and the Gulf of Mexico.

Analytical Techniques

For tract evaluations for lease sales, the MMS uses the MONTCAR program, which calculates economic values for individual tracts within a prospect. In resource evaluations, the MMS uses a different program, PRESTO (also discussed under "Economic Boundaries"). PRESTO performs multiple simulations of industry exploratory drilling and ranks economically successful drilling efforts in terms of resources discovered and probabilities of occurrence. To determine the total volume of undiscovered recoverable resources for the 1989 assessment, the MMS modified its data base to remove the economic constraints and used statistical methods to extrapolate the size and number of all potential fields above 1 Mmboe.

MMS assessors supplied the PRESTO input parameters differently from region to region. Most of the differences were related to basic play definition. For example, 34 plays were delineated in Alaska, while only 6 were delineated in the Gulf of Mexico, even though the Gulf is the most extensive and most productive OCS region. Other differences between regions were related to the economic screens applied. Major differences, however, tended to be reconciled by peer review within provinces and regions, by later MMS management review, and by a process of arriving at a consensus.

Alaska Assessment Forms

In Alaska, two basic forms, A and B (see Figures 3.3 and 3.4), were used to standardize input parameters for plays and prospects. The committee found fault with two aspects of the procedure imposed upon the evaluators.

The first aspect of the procedure that deserves review is the use of productive acreage of potential pay zones to define vertical fill-up of those zones. (Formerly, an arbitrary upper fill-up limit of 1500 feet was used.) The committee

FORM A

PROSPECT PROBABILITY OF SUCCESS
for
PRESTO PROGRAM
NATIONAL RESOURCE ASSESSMENT

Your Name(s)________________________________Date__________

Name of Province________________________________

Name of Play________________________________

Prospect Number________________________________

For each of the following probabilities, assign a number zero to one, where zero indicates no confidence, and one indicates absolute certainty.

1. Province: What is the probability that at least one play in the province contains physically recoverable resources?

   Probability of Success = ____________________

2. PLAY: What is the probability that at least one prospect in the play contains physically recoverable resources?

   Probability of Success = __________(Must be ≤ Province Probability)

3. Prospect: Assign probabilities for trap, reservoir, and geologic history.

   A. TRAP: Consider what kind of trap is this prospect, what is the probability that it exists as mapped, and what is the quality of the seal.

Trap Type	Probability of Success =
Simple Anticline	(max. 1.0)
Faulted/Truncated Anticline	(max. .75)
Fault Trap	(max. .50)
Strat Trap	(max. .25)

   B. RESERVOIR: What is the probability that the estimated minimum recovery factor and the minimum net pay exists within this prospect?

   Probability of Success = ____________________

   C. GEOLOGIC HISTORY: What is the probability that the geologic history of this prospect is favorable toward the sourcing and migration of hydrocarbons into this trap and reservoir?

   Probability of Success = ____________________

PROSPECT PROBABILITY OF SUCCESS = A x B x C = ____________________
(Must be ≤ Play Probability)

FIGURE 3.3 Form A, the first of two forms that MMS assessors used to record input parameters for PRESTO in Alaska.

FORM B

PRESTO RESOURCE ESTIMATE REPORT FORMAT

Your Name(s)__________ Date__________
Name of Province__________ Sale Number__________
Name of this Play__________
This Play Probability of Success (from Form A)__________
This Prospect Number______ Prospect Probability of Success (from Form A)__________
Water Depth for this Prospect (Feet)__________
Distance from Shore for this Prospect (Statute Miles)__________

		G/G - Supplied Distributions		
		Zone 1	Zone 2	Zone 3
Depth to Top of Zone (Feet)				
Zone Probability of Success*				
Total Closure at Spill Contour, Acres				
OPROB**________	GPROB**________	//////////	//////////	//////////
Proportional Gas Pay	Minimum			
(Fraction of Net Pay)	Most Likely			
	Maximum			
Productive Acres	Minimum			
(Defines Fill-Up)	Most Likely			
	Maximum			
Pay Thickness	Minimum			
(Net Feet)	Most Likely			
	Maximum			
Oil Recovery Factor	Minimum			
(Bbls/Acre-Foot)	Most Likely			
	Maximum			
Gas-To-Oil Ratio	Minimum			
(CF/Bbl for Dissolved	Most Likely			
Gas)	Maximum			
Gas Recovery Factor	Minimum			
(MCF/Acre-Foot for Gas	Most Likely			
Cap & Nonassociated Gas)	Maximum			
Natural Gas Liquids	Minimum			
(Bbls/MMCF for Gas Cap	Most Likely			
& Nonassociated Gas)	Maximum			

* Zone Probability of Success must be ≤ Prospect Probability of Success. For a single zone, Zone Probability = Prospect Probability.

** OPROB = Probability of all Oil; GPROB = Probability of all Gas. OPROB + GPROB must be ≤ 1.00. The computer generates a random number on the interval 0 to 1. If the number falls between 0 and OPROB, the zone contains oil only. If the number falls between 1-GPROB and 1, the zone contains gas only. Otherwise, both are present. In case the zone contains both oil and gas, the proportion of the gas pay is determined by sampling the G/G-supplied distribution.

All Oil	Oil and Gas	All Gas
Example: OPROB = 0.40	Gas Fraction Developed By PRESTO	Example: GPROB = 0.30

0.00 1.00

FIGURE 3.4 Form B, the second form that MMS assessors used to record PRESTO input parameters in Alaska.

judges that it would be more appropriate to relate vertical fill-up to vertical closure and area of closure, as well as to productive acreage. Further, in defining input parameters like vertical fill-up, the MMS should develop better guidelines to allow geoscientists' judgments to be flexible by play or province instead of imposing region-wide, rigid directives. The second problematic aspect of the procedure was the imposition of arbitrary productive acre limits, apparently to prevent estimates from becoming "too high." In the Chukchi Sea, for example, the lowest enclosing contours of the prospect included 96,000 acres. The numbers used in the resource evaluation were: minimum, 9,600 acres; most likely, 19,200 acres; maximum, 28,800 acres. Thus, acreage sizes used were 10 percent, 20 percent, and 30 percent of the acreage within the lowest enclosing contour. A potential reserve generated by the "most likely" values used for this prospect would be obtained from the following computation:

$$19{,}200 \text{ acres} \times 310 \text{ barrels/acre ft.} \times 200 \text{ ft.} = 1.190 \text{ Bbo}$$

(Two-hundred feet is the thickness of the zone.) By contrast, if a "most likely" acreage were 50 percent of the total closure, or 48,000 acres, the comparable reserve would equal 2.976 Bbo (using the same recovery factor and net pay). The committee concludes that while conservative judgments about productive acreage may be appropriate for lease sale tract evaluation purposes, they are not appropriate when applied to resource assessments designed to evaluate undiscovered oil and gas in a play.

Probabilistic Dependence

The MMS's assessment procedure, like the USGS's procedure, did not incorporate a way to address possible probabilistic dependencies between uncertain quantities. In the Atlantic offshore region, for example, the MMS used five generic play types for the 1989 assessment: plays underlying rifts (Triassic grabens), plays along the shelf edge (Jurassic carbonates), sedimentary pinch-outs seaward of the Jurassic carbonates, lower Cretaceous drapes over basement highs seaward of the Jurassic carbonates, and faulted anticlines on the continental shelf. Seismic profiles indicate that structures that are well-defined at depth "die out" and porosity increases upward from the Triassic strata. The implication is that structure, size, and porosity are negatively correlated as functions of depth. In addition, structural relief and extent appear to be correlated with depth. These empirical findings suggest that an assessment procedure in which all uncertain quantities assessed are assumed to be probabilistically independent will misrepresent an essential feature of this particular depositional environment. In particular, if structure quality is depth dependent, prospect risk must also be depth dependent. The committee recommends that the MMS modify its assessment procedure to allow the incorporation of probabilistic dependencies between uncertain geophysical quantities like structure, size, porosity, and depth.

MMS assessment procedures were based on the assumption that individual prospect outcomes are mutually independent conditional on the presence of at least one field. Careful study of the statistics of well-explored plays will clarify whether or not such an assumption is reasonable. No such studies have, to our knowledge, been completed. Because of the nature of risking, assuming independence can tend to increase the estimated risk of failure to find recoverable resources. For example, the committee reviewed a case in the Pacific OCS that we believed exhibited dependency of risk. The MMS estimated the individual chance of success (1.0 minus the percent risk of failure) to be .73 based on an assumption of complete independence. According to the committee's calculations, assuming high dependency among zones would have yielded a .85 chance of success (see Appendix A). This .12 difference in the success chance when prospect outcomes are assumed dependent instead of independent can increase the estimated recoverable resources by many millions of barrels. Thus, a tendency to prefer an assumption of independence of risk will tend to result in overly conservative resource estimates. Consequently, it is critical that the MMS (and the USGS) refine its current risk assessment practice. The committee recommends that the MMS undertake a study to determine if a pattern of dependencies is present in observed sequences of prospect drilling outcomes. The MMS should redesign assessment methods and forms to incorporate dependencies where present.

Assumptions

As discussed previously, the MMS's resource assessment method relied on many decisions regarding the selection of input parameters entered into PRESTO. Many input parameters can be determined from the general data base, experience, and reasonable projection or extrapolation. In areas with a long history of exploration and production, the range of choices for input parameters can often be well defined and fairly narrow in span. In frontier areas, however, the range of choices can be extreme and very difficult to establish. MMS geoscientists and engineers for the most part appear to have used good judgment in selecting PRESTO input parameters. Some specific instances, however, require examination and potential revision.

In Alaska, a recovery factor of 40 percent for oil was applied universally in the 1989 assessment with secondary recovery initiated at the inception of production. Because of the lack of federal OCS production history in Alaska, this assumption may well be a reasonable generality. It does, however, conflict with the USGS's assumption of a 32.3 percent recovery factor on the North Slope. The committee recommends that as the data base expands, the MMS should shift to the use of play-oriented recovery factors related to identifiable lithologies, depths, and overall reservoir characteristics.

In the Pacific OCS region, the recovery factors the MMS used for the Monterey reservoirs were based upon gross formation thickness, using a barrels-per-acre-foot recovery from analogs from productive Monterey fields. The committee advises that the MMS attempt to define more precise methods, based upon net pay, for determining recovery factors. In fairness, it should be realized that industry now uses similar gross thickness calculations or barrels-per-acre analogy for reserve estimates. Nevertheless, continued study and review could possibly improve methods for future assessments.

In Alaska, the MMS applied an arbitrary area cut-off so that prospects less than one-half a leasing block were not included. They similarly excluded prospective reservoirs with less than 100 feet of net pay and reservoirs occurring at depths of less than 3,000 feet. Prospects judged to be subeconomic (if appropriately risked) were not modelled in the conditional case. These four exclusions are assumptions based upon economic screens that the committee considers to be inappropriate in determining the oil and gas resource endowment. The committee recommends that the MMS develop methods for separating technically recoverable resource calculations from those that determine economic resources, which is difficult with current PRESTO methodology.

Economic Input

The committee generally viewed the MMS's economic parameters as reasonable based on current industry data and practices. Likewise, the MMS's assumptions regarding minimum economic field sizes appeared reasonable and well defined in the various operating areas. The determination of the MEFS for prospects in different plays, water depths, operating conditions, producing or frontier areas, and drilling depths is a major component of the PRESTO program. In Alaska, for example, in areas of difficult operating conditions, such as in the Beaufort and Chuchki Seas, minimum field sizes were appropriately escalated to account for expensive production and transportation infrastructure requirements, and separate calculations were made for "satellite" versus "stand-alone" fields.

Future assessments will include producing environments in even deeper water and possibly more remote areas and hazardous operating conditions. Future technological developments and innovative technologies for drilling and producing operations will require even more economic screening models to present the available options. Continued MMS efforts in the engineering and economic modelling and data gathering programs should be encouraged to insure continued high-quality results.

Uncertainties

In frontier areas with little or no established commercial production, the uncertainties in input parameters, as well as in resulting estimates, are very large. For example, the 1989 assessment indicated that the mean value of the Alaska OCS recoverable oil resource is 3.4 billion barrels. This single number may be misleading if not compared to the *range* of possible values, as indicated by the F95 and F5 fractiles. (The F95 fractile is the petroleum volume for which there is a 95 percent probability that the area contains more than that volume; the F5 fractile is defined analogously.) In Alaska, this range is very broad: 0.6 to 9.4 billion barrels.

The committee is concerned that the MMS's method of reporting resource values focused too much on mean values. In reporting future assessments, the MMS should emphasize more the role uncertainty plays in resource estimates and use more graphic displays to demonstrate visually the ranges of uncertainty. This type of reporting would assist the non-technical user in understanding the process and results, and possibly offset the tendency to focus only on the single number, the mean, which is now most often reported and used.

Statistical Methods

Number of Undiscovered Fields

MMS assessors obtained distributions showing the number of undiscovered fields by probabilistic sampling and economic screening of identified prospects.

The identified prospects had specific geographic locations, risk factors, physical information relating to field-size distribution, and economic screening information. In addition to identified prospects, the MMS allowed for the inclusion of unidentified prospects in each basin. The number and field-size parameters of unidentified prospects were extrapolated from the statistics of identified prospects. (As mentioned previously, it appears that unidentified prospects were included infrequently.)

The MMS assessment procedure converted a prospect to the inventory of undiscovered fields by a coin toss, with conversion probability equal to one minus the risk factor for that prospect. Prospects were risked independently of one another, given that at least one prospect in the basin would be converted. The probability that no prospect in a basin would be converted was taken to be the basin risk. This basin risk implicitly introduced a degree of dependence among prospects, discussed later.

If the prospect survived the coin toss, then a field size was assigned. The field was retained in the inventory only if its assigned size exceeded a minimum economic size screen specific to that prospect. The economic screen was set to 1 Mmboe for all prospects for "undiscovered recoverable resources" estimation (as opposed to "undiscovered economically recoverable resources" estimation). The total number of undiscovered fields in a basin thus became a random fraction of the total number of identified and unidentified prospects.

The MMS made no systematic attempt to calculate the distribution of the number of prospects that the seismic grid missed, which could be considerable even in basins with apparently extensive exploration. Likewise, the MMS did not estimate the number of prospects that were crossed by the seismic grid, but were not identified. Whenever unidentified prospects were included, their size distribution was such that they could never survive the economic screening. This is one more piece of evidence pointing to the need for the increased consideration of conceptual plays, discussed earlier in this chapter.

In contrast to the MMS, the USGS did not use concepts of specifically located and specifically described putative fields. For each defined play, the USGS generated multiple undiscovered fields by sampling repeatedly from a fixed, play-specific, field-size distribution that used a fixed economic cut off. This was the USGS's finest level of specificity; it is analogous to the MMS's use of unidentified prospects. <u>The committee recommends that the MMS method for assessing unidentified prospects be brought into line with the USGS's procedures for extrapolating discovery histories</u>.

Sizes of Undiscovered Fields

If a prospect survived the MMS risk screening and became an undiscovered field, it was assigned a field size. This field size was generated as a product of area, thickness, and recovery numbers, each randomly and independently drawn from distributions specific to that prospect.

The attempt to break down the field size into more easily described physical quantities may seem reasonable. It allowed for separate assignment of distributions for the input area, thickness, and recovery of the selected prospect. However, one then needs to put these numbers back together somehow and this was done by assuming statistical independence of the three inputs. The committee recommends that the MMS compare field-size distributions produced under the assumption that area, thickness, and recovery values are independent with empirical field-size data. This comparison will show whether independence of field-size parameters is a reasonable assumption.

MMS assessors typically specified probability distributions for quantities like thickness on prospect evaluation forms that differed somewhat from region to region. Evaluators familiar with the prospect were asked to specify minimum, most probable, and maximum values for thickness. Through an undocumented procedure, these three values were turned into one of several possible distributions. It appears that "minimum" and "maximum" were to be interpreted as the lower and upper 5 percentage points of the distribution and "most probable" as the mean value. As discussed earlier, it is not clear whether individual evaluators were aware of these interpretations.

Geologic Risk Factors

Not all zones or prospects are undiscovered fields. The MMS assessed this uncertainty through a hierarchy of risk numbers associated with the geologic hierarchy of zones, prospects, basins, and areas. The procedure can be illustrated by considering the assignment of prospect and basin risk numbers. The prospect risk is defined as one minus the probability that the prospect is an undiscovered field. The basin risk is one minus the probability that the basin contains undiscovered fields. For discussion here, assume that zones and prospects coincide.

The risk for each prospect was calculated on a work sheet that varied somewhat from one MMS region to another. The work sheet required the determination of risk factors for three geologic attributes: the presence of a trap, a reservoir, and petroleum. The three risk factors were then mechanically combined as though they were statistically independent. For example, a 50 percent risk on each risk component produces an 87.5 percent risk $[1 - (.5 \times .5 \times .5) = .875]$ for the prospect (i.e., an 87.5 percent chance that the prospect contains no petroleum). (Some MMS regions used a more detailed risk analysis work sheet in which each of the three basic components of prospect risk were broken down further.)

The assignment of a basin risk factor was constrained so that it was never less than the combined risk of the component prospects considered as statistically independent trials. This created a degree of positive correlation among the outcomes of the prospects within a basin, which is good in principle. In typical situations with a large number of prospects in a basin, the lower risk limit for the basin is essentially zero; therefore, even small assigned basin risk numbers will introduce substantial correlation among the outcomes of prospects. The basin risk was also constrained so that it did not exceed the minimum risk for any component prospect.

The problem with this method of assigning risk numbers is that it is susceptible to misinterpretation on the part of the dispersed group of risk evaluators. In assessing a specific prospect risk, one must to some extent anticipate basin-level and higher-level risk assessments. It could be argued that it is more natural to assign a prospect risk number *conditionally* on resources being present in the basin. In particular, the reservoir and hydrocarbon risk factors used in a prospect risk calculation should in many cases reflect properties of groups of prospects or even of the whole basin. The committee recommends that MMS assessors separate more clearly risk factors that are shared versus risk factors that behave independently.

The risk numbers assigned to basins, and especially to areas, are extremely critical to the overall resource assessment. Great care in assigning prospect-level risks can be vitiated by misunderstood or carelessly assigned risk factors at higher levels. Detailed work sheets were not used for these higher level assessments but perhaps should have been to insure consistency.

It is the committee's impression that the MMS set area-level risks at insupportably high levels. Many areas in the Atlantic and Alaska regions had area risks in the 90- to 98-percent range prior to any economic screening. A risk of 95 percent for an area implies that only one out of 20 similar areas would contain even a single undiscovered field of any size.

One of the difficulties in assigning risk numbers to higher levels of the risking hierarchy is the avoidance of confounding the numbers with economic screening. The committee's impression of exaggerated area-level risk numbers may derive from unintended but implied economic considerations. The committee recommends that the MMS explicitly and clearly separate economic considerations from basin- and area-level risk assignments to avoid unintended double discounting.

The assignment of risk numbers, particularly at the higher levels of the risking hierarchy, is necessarily a nebulous task, but it nevertheless exerts a strong influence on the final undiscovered resource estimates. Therefore, the committee recommends that the MMS attempt to validate or calibrate risk numbers to increase the credibility of its assessments. The MMS should offer some assurance that assigned risk numbers are not systematically too high or too low, are not confounded with implied economic screening, and are appropriate to the hierarchical nature of risking. Providing such assurance should involve experimental work that compares multiple risk numbers across individuals and geographic units and that employs whatever historical information is available.

It is inevitable that different individuals with comparable knowledge will assign different risk numbers in the same situation. The MMS should consult statistical literature on methods for combining diverse expert opinion. The committee believes that diversity of opinion on risk numbers should be propagated to the final reported range of resource estimates.

Aggregation

The MMS aggregated undiscovered resources across zones, prospects, and basins by using the assigned hierarchical geologic risk numbers to determine probabilistically which prospects (zones) contain undiscovered petroleum. The sizes of these undiscovered fields were also probabilistically assigned from prospect-specific size distributions, as described above. For the economic resource estimate, some or all of the converted prospects did not survive the hierarchical economic screening, which was prospect-, basin-, and area-specific.

The net result was a realization of a probabilistically selected collection of undiscovered fields for a given area. The total of the selected field sizes was computed. This process of probabilistically estimating an area aggregate was repeated many times to generate a frequency distribution of area aggregate resources. Finally, a number of zeroes were added to the frequency distribution; the number of zeros corresponded to the area-level risk number. For example, if there were 1,000 simulated area aggregates and the area risk was 90 percent, then 900 zeroes were added to the distribution of aggregates. For each area of each region, the MMS reported the mean value, together with the fifth and ninety-fifth percentiles, of this combined aggregate distribution.

In the appendices of the DOI's assessment report, the MMS published "conditional" estimates of area resources. Conditional estimates represent an area's total resource estimate without the zeroes added for the area risk. In other words, conditional estimates assume that there is at least one petroleum-containing field in the area. Of course, these "conditional" resource estimates still contain substantial amounts of discounting for the geologic and economic risks applied at lower levels of the risking hierarchy. Assessment users may be confused by the meaning of conditional estimates. In future assessment reports, the MMS should provide a better interpretation of the reported conditional estimates. The MMS could compare the conditional estimates with past results and explain the causes of significant differences and implications for future production prospects.

In addition to reporting resource estimates for each area, the MMS reported estimates for each region. Although the assessment report does not document the method for aggregating area resources to obtain regional estimates, it appears that area-specific distributions were repeatedly sampled to generate regional distributions and that areas were assumed to be statistically independent. As mentioned in the evaluation of USGS assessment methods, the MMS's assumption that areas are statistically independent conflicts with the USGS's assumption of complete conditional dependence among plays. For future assessments, the committee recommends that the USGS and MMS standardize their procedures for aggregating resources.

REFERENCES

Adelman, M. A., J. C. Houghton, G. M. Kaufman, and M. B. Zimmerman. 1983. Energy Resources in an Uncertain Future: Coal, Gas, Oil, and Uranium Supply Forecasting. Cambridge, Massachusetts: Ballinger Publishing Company.

Association of American State Geologists. 1988. Review of Geologic Information Utilized by the U.S. Geological Survey and Minerals Management Service in their Assessment of U.S. Undiscovered, Conventionally Recoverable Oil and Gas Resources. Tulsa, Oklahoma: Association of American State Geologists.

Attanasi, E. D. 1988. Minimum commercially developable field sizes for onshore and state offshore regions of the United States. Pp. 90-117 in National Assessment of Undiscovered Conventional Oil and Gas Resources. U.S. Department of the Interior Open-File Report 88-373, Washington, D.C. Photocopy.

Crovelli, R. A., R. F. Mast, G. L. Dolton, and R. H. Balay. 1988. Assessment methodology for estimation of undiscovered petroleum resources in play analysis of the United States and aggregation methods. Pp. 30-48 in National Assessment of Undiscovered Conventional Oil and Gas Resources. U.S. Department of the Interior Open-File Report 88-373, Washington, D.C. Photocopy.

Energy Information Administration. 1990. U.S. Oil and Gas Reserves by Year of Field Discovery. Report CIA-0534. Washington, D.C.: U.S. Department of Energy.

Energy, Mines and Resources, Canada. 1977. Oil and Gas Resources of Canada, 1976. Report EP77-1. Ottawa, Canada.

Fisher, W. L. and W. E. Galloway. 1983. Potential for Additional Oil Recovery in Texas. Circular 83-2. Austin, Texas: University of Texas Bureau of Economic Geology.

Houghton, J. C., G. L. Dolton, R. F. Mast, C. D. Masters, and D. H. Root. 1988. The estimation procedure of field size distributions for the U.S. Geological Survey's national oil and gas resource assessment. pp. 56-72 in National Assessment of Undiscovered Conventional Oil and Gas Resources. U.S. Department of the Interior Open-File Report 88-373, Washington, D.C. Photocopy.

Minerals Management Service. 1985. Estimates of Undiscovered, Economically Recoverable Oil and Gas Resources for the Outer Continental Shelf as of July 1984. OCS Report MMS 85-0012. Washington, D.C.: U.S. Department of the Interior.

Morton, R. A. and D. Nummedal, eds. 1989. Shelf Sedimentation, Shelf Sequences and Related Hydrocarbon Accumulation: Proceedings, Seventh Annual Research Conference, Gulf Coast Section of the Society of Economic Paleontologists and Mineralogists. Austin, Texas: Earth Enterprises, Inc.

National Research Council. 1988. Energy-Related Research in the U.S. Geological Survey. Washington, D.C.: National Academy Press.

Podruski, J. A., J. E. Barclay, A. P. Hamblin, P. J. Lee, K. G. Osadetz, R. M. Procter, and G.C. Taylor. 1988. Conventional Oil Resources of Western Canada. Paper 87-26. Ottawa, Canada.

Tyler, N., W. E. Galloway, C. M. Garrett, Jr., and T. E. Ewing. 1984. Oil Accumulation, Production Characteristics, and Targets for Additional Recovery in Major Oil Reservoirs of Texas. Circular 84-2. Austin, Texas: University of Texas Bureau of Economic Geology.

U.S. Department of the Interior, Geological Survey and Minerals Management Service. 1988. National Assessment of Undiscovered Conventional Oil and Gas Resources. Open-file Report 88-373, Washington, D.C. Photocopy.

U.S. Department of the Interior, Geological Survey and Minerals Management Service. 1989. Estimates of Undiscovered Conventional Oil and Gas Resources in the United States—A part of the Nation's Energy Endowment. Washington, D.C.: U.S. Government Printing Office.

U.S. Geological Survey. 1975. Geological Estimates of Undiscovered Recoverable Oil and Gas Resources in the United States. Circular 725. Washington, D.C.: Department of the Interior.

White, D. A. 1980. Assessing oil and gas plays in facies-cycle wedges. American Association of Petroleum Geologists, Bulletin 64: 1158-1178.

4
RECOMMENDATIONS

In preparing this report, the committee undertook to provide the USGS and the MMS with concrete suggestions for improving their future estimates of undiscovered oil and gas volumes. Such estimates may be more important now than at any time since the energy crises of the 1970s. With the recent upheaval in the Persian Gulf, domestic pressure to decrease the nation's reliance on imported oil and to produce more oil from U.S. sources will probably escalate. Reliable resource assessments are critical for gauging the extent to which the United States can depend on undiscovered domestic petroleum for its future energy supply.

Estimating how much oil and gas remains to be discovered is necessarily an inexact process. Without actually drilling, one cannot know precisely what volume of petroleum a prospective reservoir contains. Nevertheless, new assessment methods developed since the early 1970s have the ability to increase the reliability of resource assessments if properly employed. These new methods, however, have also increased the complexity of assessment procedures. Today, conducting a successful assessment requires expertise not only in petroleum geology, but also in statistics. Because of the complex, interdisciplinary nature of resource assessments, outside review of assessment procedures is critical to ensuring that evaluators have correctly applied the most up-to-date geological and statistical appraisal methods.

This chapter summarizes the committee's most important recommendations related to the 1989 assessment. This summary does not cover every finding and recommendation from Chapter 3. Instead, it focuses on the committee's key conclusions—those most important to the assessment's overall quality. For this summary, the committee has divided its recommendations into five areas:

assessment boundaries, USGS and MMS management, geological approaches in data base use and play analysis, statistical methods, and assessment results.

ASSESSMENT BOUNDARIES

The committee found that the assessment's limitation to "conventional" resources was unnecessary in some cases. While the types of *oil* the assessors classified as unconventional generally contribute little to the nation's energy supply, some types of *natural gas* labelled unconventional represent a significant fraction of domestic production. For example, natural gas produced from low-permeability sandstones, fractured shale, and coal beds amounts to about 1.5 to 2 trillion cubic feet per year. The failure to include estimates of unconventional natural gas in addition to the estimates of conventional natural gas obscures a significant portion of the potential domestic energy supply.

The committee also found that the assessment's limitation to recoverable resources made the estimates unnecessarily sensitive to future changes in recovery technology. There is a continual flux of resources from the "unrecoverable" category to the "recoverable" category, as advances in reservoir characterization, drilling, and completion increase the amount of in-place petroleum that can be produced. Thus, a more thorough assessment would include estimates of in-place resources in each play, which could be used to judge the potential for new recovery technology to increase petroleum production.

Recommendation: The DOI should include estimates of natural gas from unconventional sources (separate from the estimates of conventional natural gas) in future resource assessments.

Recommendation: The DOI should include estimates of in-place resources in future assessments. Assessors should estimate each play's in-place resource volume first, and then calculate the recoverable resource volume by applying a recovery factor to the in-place value.

USGS AND MMS MANAGEMENT

To be done well, national-scale resource assessments demand a major commitment of funding and professional expertise. Credible resource assessments also require *continuity* of effort. Continued funding and effort are necessary to improve the coverage and quality of data bases and to translate the newest theoretical research on assessment methods into well-tested modifications of actual assessment procedures. Sporadic application of intensive effort just prior to a national resource assessment is unlikely to result in an outcome that achieves a uniform standard of excellence in all dimensions by which such assessments are judged. The soundness of assessment procedures, the caliber of their execution, and the credibility of their results are thus linked to the fashion in which the program for accomplishing the assessment is managed.

The committee found that, as of the time of the 1989 assessment, the USGS management had provided insufficient manpower, funds, and incentives to carry out a national oil and gas assessment at a uniform level of excellence in all dimensions. The USGS, unlike the MMS, does not maintain a permanent resource assessment group. The absence of a permanent assessment group created the following problems:

- a lack of staff experience in resource assessments and insufficient knowledge of regional geology in some provinces;
- a distribution of personnel among the provinces that did not correlate with the provinces' potential for containing undiscovered oil and gas or with the amount of data available for analysis, with a concentration of attention on the Rocky Mountains, California, and Alaska and a comparatively low level of effort in the Gulf of Mexico, Midcontinent, and Illinois Basin areas;
- the lack of a clear explanation of how the 1989 assessment methodology compared to methodologies used in previous assessments;
- an absence of a clearly defined assessment procedure that was unambiguously understood by all members of the assessment team;
- insufficient statistical support for data analysis; and
- a lack of provision for improving assessment methodologies or testing the sensitivity of the methodologies to critical input parameters.

The MMS avoided some of these problems by maintaining a permanent resource assessment group. A permanent assessment group is a necessity for the MMS, since it is required to report undiscovered oil and gas estimates for the Outer Continental Shelf to Congress every two years. Nevertheless, the committee is concerned that funding cutbacks caused by moratoria on lease sales may cause the MMS to reduce personnel levels. As documented by the problems the USGS encountered because of its lack of a permanent resource assessment team, cutbacks at the MMS could have a negative impact on future assessments.

Recommendation: The USGS should establish a group of specialists within its offices to design and implement on an ongoing basis a program for improving oil and gas assessment methodologies. In particular, this permanent assessment group should emphasize data validation, training of geologists in assessment methods, and more aggressive use of modern statistical methods in the assessment process.

Recommendation: Managers at the USGS and the MMS should establish complete, continued cooperation between the assessment groups at the two agencies.

GEOLOGICAL APPROACHES IN DATA BASE USE AND PLAY ANALYSIS

The committee examined two aspects of the assessment's geological framework: (1) the adequacy of the geological data bases available to the assessors and (2) the formulation and analysis of plays.

Geological Data Bases

Geological data bases provide the foundations for resource assessments. Without good data, the most advanced statistical methods are powerless. The committee found weaknesses in the data bases used by both the USGS and the MMS.

After reviewing the USGS's data, the committee found that the USGS lacked adequate seismic data for the lower 48 states and also for Alaska's North Slope. In addition, the committee found that the information compiled by USGS assessors into basin reports varied in scope and quality, and that the preparation of this information did not occur for all basins as an established, early part of the assessment process.

From a review of the MMS's data, the committee concluded that while the MMS's seismic data base was relatively complete, there were gaps in the MMS's geochemical data. The committee also concluded that the MMS did not make adequate use of potential data from key industry exploratory wells in Alaskan offshore basins and in the Atlantic's Baltimore Canyon.

Recommendation: Both the USGS and the MMS should conduct a data audit. The audit would have three purposes: (1) to evaluate the accuracy and completeness of the data; (2) to identify areas where the data base requires improvements; and (3) to provide explicit measures of the data quality to assessment users.

Recommendation: Based on the results of the data evaluation, both the USGS and the MMS should attempt to expand the available data for areas where their present data bases are incomplete. Better use of existing data bases should precede the creation of extensive new data bases. The agencies should seek data from outside sources like state geological surveys, state regulatory agencies, other federal agencies, and the private sector. For example, it is possible that the USGS could obtain some degree of access to proprietary seismic lines from industry in key areas.

Recommendation: To expand its geochemical data base, the MMS should ensure that full geochemical evaluations are conducted for wells drilled in offshore areas where existing data are inadequate.

Play Formulation and Analysis

Fundamental to modern oil and gas assessment methods is the proper grouping of prospects into plays. When plays are defined improperly—when diverse depositional systems are combined in one play—the result is play mixing. Play mixing can result in inaccurate statistical characterization of the play's prospects. Inaccurate statistics, in turn, may lead to an inaccurate assessment of the play's potential to contain undiscovered resources. The most likely result of play mixing is the *underestimation* of undiscovered resource volumes.

Also important in play analysis is the consideration of conceptual plays: those that do not contain discoveries or reserves, but that geological analysis indicates may exist. Analyzing conceptual plays is especially important for assessing natural gas, because natural gas exploration is less mature than oil exploration. Like play mixing, the failure to consider conceptual plays results in an underestimation of resource volumes.

The committee found that the USGS created some excessively large and lithologically diverse plays, leaving open the potential for play mixing. Large plays were most notable in the Gulf Coast and mixed plays were most notable in the Permian Basin, both important provinces. The committee also found that the USGS did not provide adequately for conceptual plays. The evidence of play mixing and incomplete consideration of conceptual plays indicates that the 1989 resource estimates may have been too low.

The USGS and the MMS used different procedures for delineating plays. The MMS's plays consisted either of a summation of prospects or a set of broad stratigraphic subdivisions, as in the Pacific Coast region. Thus, the committee concluded that the MMS's play delineation procedure created even more potential for play mixing and the consequent underestimation of resource volumes than the USGS's procedure. Part of the reason for the MMS's imprecision in play delineation was a lack of exploration and thus a shortage of drilling data everywhere except in the Gulf of Mexico. However, even in the Gulf, plays included a wide range of geologic settings, lithologies, stratigraphies, depositional systems, and structural characteristics, making the mixing of plays unavoidable.

As with the USGS, the committee found that the manner in which the MMS defined conceptual plays did not result in the proper identification and categorization of such plays and thus missed their potential contribution to the overall resource endowment. In Alaskan waters, the limited number of conceptual plays was inconsistent with the region's early development stage—a stage at which conceptual plays should be at a maximum.

Recommendation: The USGS should analyze play content to determine the impact of play formulation—especially formulation of plays with diverse depositional systems—on the resource volumes reported in the 1989 assessment. In future assessments, the USGS should avoid creating excessively large or geologically diverse plays.

Recommendation: The MMS should define plays more carefully to avoid the mixing of diverse geological and reservoir engineering characteristics in one play. Plays should not be simply summations of prospects. The MMS should recognize that the availability of an extensive seismic data base could lead toward plays excessively dependent on recognition of structural traps. Once it has formulated plays appropriately, the MMS should institute testing to ensure that play mixing does not significantly alter resource estimates.

Recommendation: Both the USGS and the MMS should incorporate more conceptual plays in future assessments. Analyzing conceptual plays may require that the agencies develop assessment techniques different from those used for known plays.

STATISTICAL METHODS

Statistical methods, modulated by expert judgement, play a central role in translating geological, geochemical, and geophysical data into projections of undiscovered oil and gas resources. A scientifically credible assessment program must be flexible enough to adapt statistical methods to changes in the quantity and scope of available data. Consequently, a credible assessment program should vigorously embrace two data analytic activities: data validation and model validation. Equally important, a successful assessment program must train participating geologists and geophysicists in statistical methods.

The USGS's attempt to incorporate play analysis into the 1989 assessment, despite the initial difficulties it experienced in implementing this concept, is commendable. However, the program of statistical work leading up to the assessment had four significant weaknesses:

1. The USGS's method for extrapolating discovery data to determine the number and size distributions of undiscovered fields in a play relied principally on subjective judgment. Extrapolations were based on observations about the way the shape of the undiscovered field-size distribution changes with time—observations obtained by fitting sizes of discovered fields to truncated shifted Pareto distributions (see Chapter 2, under Assessment Methods, for an explanation of this procedure). Where a moderate to large discovery history is available, more up-to-date, objective discovery-process models could have provided projections of undiscovered oil and gas in plays unencumbered by judgmental uncertainties that accompany subjective extrapolations. Objective discovery-process models could also have provided a benchmark for the use of expert opinion.

2. With the exception of reserve growth studies, objective statistical methods were not applied as aids in analyzing and interpreting data and plays. For example, the committee uncovered little evidence that the USGS used currently available statistical methods for partitioning data into descriptively homogeneous subsets (here, subsets composed of distinct individual play data).

3. No formal statistical analysis of data was employed to support or reject fundamental assumptions about probabilistic independence of play attributes like "adequate timing," "existence of migration paths," and "existence of reservoir." The USGS reported no tests to determine how sensitive its undiscovered oil and gas projections are to this independence assumption.

4. The USGS assumed incorrectly that the *number* and *size distribution* of fields remaining undiscovered are statistically independent. Models of the petroleum discovery history show that the sizes of undiscovered fields in a play are somewhat related to the number of undiscovered fields in the play. As the number of undiscovered fields decreases (i.e., as more fields are discovered), the average size of remaining fields decreases, because large fields tend to be discovered first.

These four problems with the USGS's statistical work signal a major gap: the absence of a well-structured portfolio of statistical activities designed to support oil and gas resource assessments.

Though the MMS has a permanent staff devoted to resource assessments, the committee pinpointed weaknesses in its statistical methods, too. Like the USGS, the MMS assumed variables were probabilistically independent more often than was justified. In delineating plays in the Atlantic offshore, for example, the MMS assumed that structure, size, porosity, and depth are independent. In calculating sizes of prospective fields, it assumed area, thickness, and recovery are independent parameters. And in simulating prospect drilling, it assumed that individual prospect outcomes are mutually independent conditional on the presence of at least one field. Without conducting statistical tests to verify these assumptions of independence, it is impossible to conclude definitively whether the assumptions are valid.

In addition to overlooking these possible probabilistic dependencies, the MMS may have unintentionally imposed economic constraints on calculations of technically recoverable resources. For example, in Alaska the MMS excluded from its technically recoverable resource estimate prospects that were smaller than one-half a leasing block. This exclusion contains an implicit economic assumption: that prospects smaller than one-half a leasing block are too small to yield profits. When explicit economic screens are applied to technically recoverable resource volumes that were calculated with implicit economic assumptions, the result may be unintended double discounting. Such double discounting may have extended to the MMS's calculation of area risk levels (i.e., the chance that a given area contains petroleum). The committee judged that area-level risks in many regions in the Atlantic and Alaska were too high, possibly because of unintended but implied economic constraints. In these areas, risk was as high as 98 percent prior to any economic screening (i.e, the MMS assumed there was only a 2 percent chance that the area contained petroleum).

In addition to the limitations of the USGS's and MMS's statistical methods, the committee found problems with the way expert opinion was guided and articulated. In both agencies, the interpretation of important probability terms varied among geologists. For example, different geologists defined terms such as "minimum" and "maximum" possible values and ".05 fractiles" and ".95 fractiles" differently. Also, both agencies lacked standards to guide assessors in assigning play risk (the chance that a play contains petroleum). The lack of such standards leaves open the question of whether the risk numbers were systematically too high or too low. Finally, in both agencies the diversity of opinion among geologists within groups assessing play and province resources was masked by presenting only numbers that represented a consensus opinion.

Recommendation: The USGS should consider the use of objective discovery-process models wherever possible, taking into account the necessity to incorporate within these models the effect of new ideas and technologies on the discovery process.

Recommendation: To avoid unintended double discounting, the MMS should develop methods for separating technically recoverable resource calculations from those that determine the volume of economically recoverable resources.

Recommendation: Both the USGS and the MMS should conduct statistical studies of risk factors, field-size distributions, prospect drilling outcomes, and other play attributes to determine whether assumptions of probabilistic independence are justifiable.

Recommendation: Both the USGS and the MMS should consider ways to carry diversity of opinion through to the final resource estimates. For example, the agencies could report outlier opinions that, while masked by the aggregated results, would lead to significantly different resource estimates.

Recommendation: Both the USGS and the MMS should develop explicit standards for estimating the risk that a play contains petroleum. These standards should be tied to qualitative descriptions of source, migration, reservoir, and timing. The agencies should train their assessment staffs in the use of these standards so that different assessors use a common frame of reference when assigning risk numbers. The standards should be designed so that they can be checked against the available geologic data and replicated by other assessors.

Recommendation: Both the USGS and the MMS should periodically train their oil and gas geologists in subjective probability assessment. The training should be more extensive than a "short course" just prior to the next national assessment and should be focused on real case histories.

ASSESSMENT RESULTS

The results of MMS and USGS resource assessments have tended to be inadequately reported by the media and poorly understood and misinterpreted by key users, including members of Congress. Although some misunderstanding may be inevitable given the desire of the media and some users to deal with simple numbers, some measure of responsibility for the overall level of misunderstanding falls on the assessment team and report writers.

In particular, the committee concluded that the assessment report did not state clearly to the reader the degree of uncertainty in the various steps of interpretation, extrapolation, and conclusion from a limited data base of resource characteristics. It is critical that those preparing the resource estimates clearly communicate the uncertainties inherent in their calculations. Given the magnitude of these uncertainties, the committee is concerned that the method of reporting resource estimates focuses too much on point estimates of undiscovered oil and gas, such as the mean. Emphasizing to users the whole range of potential resource values—not just a point estimate—is especially important in Alaska because of the limited data available there and because Alaska contains such a potentially large share of the undiscovered resource base (26.9 percent of technically recoverable undiscovered oil and 22.7 percent of economically recoverable undiscovered oil, calculated with mean values).

Recommendation: In reporting future assessments, both the MMS and the USGS should place more emphasis on the range of uncertainty in their resource estimates. For example, the agencies could create more graphic displays to demonstrate visually the ranges of uncertainty.

Recommendation: The USGS and the MMS should take special care to insure that assessment users understand the relative role that undiscovered resources play in the resource base. The agencies should consider explaining and emphasizing the relative share of the undiscovered resource base as a source of reserve additions.

Incorporating the committee's recommendations in future resource assessments is likely to result in significant changes in the estimates of undiscovered oil and gas. The committee believes the overall impact will be to increase both the point estimates and the breadth of the range of estimated resource volumes. Problems with geological and statistical methods (such as mixing diverse petroleum reservoirs in plays and overlooking possible probabilistic dependencies between variables) and with the assessment boundaries (such as the exclusion of important natural gas sources) point to the conclusion that the assessment understated undiscovered resource volumes.

Implementing the recommendations requires allocation of resources over an extended time period. Especially at the USGS, the size and variable quality of data bases and limits placed on manpower dedicated to resource assessment have constrained the ability to attack resource assessment methodological problems with the vigor they deserve. Nonetheless, unless concerted effort is made to update assessment methodologies, the next national assessment may raise the same questions as the most recent assessment. By carefully considering the recommendations this report offers, the DOI can improve the credibility of future assessments.

APPENDIXES

Appendix A

Evaluation of the Hydrocarbon Resource Estimates for the Offshore Areas of Northern and Southern California and Florida South of 26° Latitude

Committee on Undiscovered Oil and Gas Resources
Board on Earth Sciences and Resources
Commission on Physical Sciences, Mathematics, and Resources
National Research Council

National Academy Press
Washington, D.C. 1989

CONTENTS

EXECUTIVE SUMMARY

The Committee on Undiscovered Oil and Gas Resources was asked to review and evaluate (1) the methods and procedures used by the Minerals Management Service (MMS) to estimate oil and gas resources for three outer continental shelf (OCS) areas and (2) the adequacy of the information for producing credible estimates for those areas. The committee established two panels to study these matters, one for the two areas of northern and southern California (lease sales 91 and 95, respectively) and the other for the South Florida Basin (lease sale 116, part 2). In working sessions at the MMS field offices where the information is maintained, the panels examined representative information, which is proprietary, pertinent to the OCS area and reviewed the methods and procedures applied to the information by the professional staff that conducted the resource estimate.

The panel that conducted the reviews of the northern and southern California areas finds that the MMS has an extensive and adequate data base for the two areas from which to develop resource estimates. The professional staff is judged to be highly qualified, and their methods and procedures in the assessment process appear to be generally appropriate. In the panel's opinion the current resource estimates are very likely conservative because of incomplete consideration of potential stratigraphic accumulations and other conceptual plays not detectable by seismic methods and because of MMS risking techniques and economic screening.

The panel was not able to evaluate directly the current MMS resource estimate for the South Florida Basin because the estimate was not available at the time the review was conducted. However, the panel examined the quantity and quality of available data and reviewed the methods and procedures employed in the resource estimation process.

Because no wells have been drilled in the basin, there is no firm basis for the determination of velocities or other stratigraphic control of the seismic data, no direct evidence of rock units that could serve as hydrocarbon reservoirs, and no direct method of recognizing potential hydrocarbon source rocks. Consequently, it is reasonable to expect large uncertainties in the resource estimate. However, good use is made of well data in adjacent areas, and the seismic lines, although sparse in parts of the basin, are of very high quality. The methods and procedures employed by the staff in extrapolating from areas with information to areas with little data are entirely appropriate for the level of information available for the basin.

The Committee on Undiscovered Oil and Gas Resources (the parent committee for the review of the three lease areas that are the subject of this report) is conducting a review of the methodologies and procedures used by MMS, and the information available to the agency in producing hydrocarbon resource estimates for all offshore areas of the United States. At the time of the preparation of the panel reports, the committee had not reached conclusions concerning some aspects of the methodologies and procedures that influence the hydrocarbon resource estimates produced by MMS. Therefore, the members of the two panels (who are members of the parent committee as well) conducted the review of these three lease areas without considering such issues of methodologies and procedures. The committee concurred with this approach, and will provide further elaboration in its report scheduled for release in March 1990.

INTRODUCTION

In his budget address to Congress on February 9, 1989, President Bush announced the establishment of a cabinet-level task force to review environmental concerns associated with outer continental shelf (OCS) oil and gas activities in three OCS lease areas: southwestern Florida (sale 116, part 2), southern California (sale 95), and northern California (sale 91). The National Research Council was asked to provide the task force with a review of scientific and technical information about environmental concerns and petroleum resources.

Committees in two boards of the National Research Council were already involved in relevant studies. The Committee on Undiscovered Oil and Gas Resources in the Board on Earth Sciences and Resources was reviewing methods of estimating onshore and offshore undiscovered hydrocarbon resources. The Committee to Review the Outer Continental Shelf Environmental Studies Program in the Board on Environmental Studies and Toxicology was reviewing the Environmental Studies Program (ESP) of the Minerals Management Service (MMS). On the basis of their experience with these nationwide reviews, the committees undertook to review the information pertaining to the three sale areas of concern. This report is from the Committee on Undiscovered Oil and Gas Resources.

To address the request, two panels of the committee were established: one panel was assigned the task of reviewing the Minerals Management Service (MMS) estimates of undiscovered hydrocarbon potential of the northern and southern California portions of the OCS (lease sales 91 and 95, respectively); the second panel was assigned to review the estimates for the southwestern Florida portion of the OCS (lease sale 116, part 2).

This report is the product of the combined efforts of the two panels.

The charge to each of the panels was to do the following:

1. Review and evaluate methodologies of estimates by MMS, and others, regarding the quantity and chemical composition of potential hydrocarbon resources in the OCS area(s).

2. Assess the adequacy and reliability of the existing scientific and technical information to make the following determinations in each subject and area under consideration; specifically (a) what is known plus reasonable extrapolation accompanied by an expression of the error or uncertainty, (b) what information is missing and the reasons why (e.g., difficulty of measurement, confounding of data, lack of theory, or insufficient time), (c) what information could be obtained with reasonable increments of investigative resources (e.g., personnel, financial support, facilities, and time), and (d) what most needs to be known to support decision making in a factual basis comparable to other similar problems?

The findings presented herein are to be considered preliminary and subject to review and possible modification by the committee in its forthcoming report.

NORTHERN AND SOUTHERN CALIFORNIA OUTER CONTINENTAL SHELF AREAS

In order to evaluate the data and methodology used in estimating oil and gas resources in the northern and southern California OCS, the panel held a three-day meeting on June 26-28, 1989, at the regional office of MMS in Los Angeles. During the panel meetings with MMS, a thorough, detailed, in-depth review of the entire scope of lease sales evaluation procedures and resource assessment processes was conducted. Los Angeles MMS regional supervisor, Robert Paul, and his professional staff presented every aspect of their methodology to the panel with complete cooperation, openness, and constructive attitudes, and full access to data, thought processes, and systems applications was given. At least 12 MMS personnel participated in the review.

Evaluation of Methodology

The panel reviewed the entire process by which economic evaluations for lease sales are derived on a tract basis and by which oil and gas resources are calculated. Because of the extensive seismic data base available to MMS in both the northern and the southern California lease sale areas, structural prospect mapping is the primary focus for identifying undiscovered resources of oil and gas. The same method used to evaluate prospects for tract economics is used in the resource assessment process.

In the MMS Pacific Region a threefold, stratigraphically subdivided play definition is used: i.e., pre-Monterey, Monterey, and post-Monterey. Undiscovered resources are aggregated from prospect to basin to area to region. This process is consistent within the region, but does not conform to "play definition and analysis" as used both offshore and onshore elsewhere in the National Assessment.

In both the MONTCAR (tract evaluation) and the PRESTO (resource assessment) programs, critical aspects lie in the methods by which risk is assessed with respect to zones, prospects, basins, and areas. Definitions and application of probabilistic dependence and independence in risk analysis may significantly alter resource estimates. For example, in one case where four zones were evaluated by individual chance of success (1 minus percent risk) to arrive at a prospect chance of success, the zones were assumed to be independent, an assumption that the panel judged to be unwarranted and which resulted in a prospect chance of success of 0.73. If the zones had been judged to be dependent, the result would have been 0.85 (zone chance factors used were 0.98, 0.97, 0.90, and 0.85) Because the chance of success is a direct multiplier in the operation which determines prospect reserves, significant reserve reduction would occur in multimillion barrel prospects. The significance of the probabilistic dependence/independence assumptions are currently under review and will be addressed in the committee's report.

It is apparent to the panel, though that current practice may result in conservative estimates of undiscovered resources in both areas. A substantial effort to improve current risk assessment practice is warranted. First, statistical studies of well data to determine the empirical nature of zone risk dependencies should be undertaken. Few have been done. Second, the results of such studies should be incorporated in MMS risk assessment procedures.

Using all available subsurface well data and extensive seismic reflection data, MMS personnel do attempt to define, map and estimate undiscovered resources for "postulated" or "unidentified" plays. These include stratigraphic traps (e.g., pinchouts or truncations), complex structures (e.g., thrust faults), diagenetic alterations (e.g., opal-chert), and statistical projections of unmapped structural closures in areas with widely spaced seismic grids.

Inherent difficulties in identifying, justifying, and quantifying unidentified prospects may result in underestimating undiscovered resources, particularly in the more remote frontier areas such as the Eel River Basin of the northern California OCS area and the outer banks and basins of the Southern California OCS area. For example, in the northern California Eel River Basin, the entire "postulated" resource was allocated to an estimated 4 "unidentified" structural prospects as compared to 92 "identified" structural prospects. In the entire basin the unidentified category yielded a risked total resource of only 5 million barrels of oil as compared to a total of 159 million barrels of oil for the identified category. This process does not included possible stratigraphic or other types of conceptual traps known to occur in other California basins.

With regard to the chemical composition of potential hydrocarbon resources, MMS has utilized available data sources, including public as well as purchased proprietary geochemical analyses of specific wells, surface sections, bottom samples, basins, and areas. Engineering data from field and production history are effectively applied by direct projection or by analog comparisons.

Determination of total basin generative capacity as a means of estimating the quantity of hydrocarbons is not used except by analogy based upon onshore basin comparisons.

Focus upon the Monterey Formation as the dominant source (and reservoir) may have the effect of downplaying other potential hydrocarbon source beds and thus tend to understate their contribution to postulated or unidentified plays in resource assessment processes. Nevertheless, the MMS methodology in characterizing the types and compositions of hydrocarbons in potential accumulations seems to the panel to be a fundamentally sound state-of-the-art application of geochemical data and principles.

At this stage in the committee's review of the MMS methodology to estimate quantity and chemical composition of potential hydrocarbon resources, the assessments of the northern California area and the southern California area seem to be adequate for economic tract evaluation decision making. The panel considers that the process appears to develop a conservative estimate through application of multiple risking approaches and economic screening. If this is sustained the committee's final report, the MMS resource estimation efforts will probably be found to understate the undiscovered resource base that could yield potentially recoverable reserves.[1]

Conclusions

In reviewing what is known, it is apparent that MMS has an unequaled, extensive data base of geological, geophysical, engineering, and production information on the southern and northern California OCS. The panel views this

---

[1] The term "recoverable reserves" is used herein in its standard connotation of referring to those identified hydrocarbons that are recoverable by all existing methods except for enhanced oil/gas recovery.

data base as adequate for making sound estimates of the hydrocarbon resources in these areas. The staff is highly qualified and applies currently accepted, state-of-the-art methods, to their analyses and interpretations. Continued study, review of methods, and sensitivity testing of results of risking and dependence versus independence could benefit the process of resource assessment.

A critical element with respect to missing information lies in the inability of MMS (or USGS) to obtain the use of certain confidential data concerning state of California waters in the 3-mile zone. The California State Lands Commission retains control over the data, which results in a critical gap in the transition area of projection of interpretive trends from onshore to offshore.

Although MMS has extensive seismic grid coverage, especially in the productive and more competitive basins and areas, northern portions of the Eel River Basin of northern California and the western extremities of the southern California outer banks and basins could be better understood with a denser seismic grid.

In order to enhance decision making based upon the relative values and significance of potential resource development, it is important to utilize the best possible knowledge and understanding of these resources. More emphasis on conceptual geologic plays or prospects could improve the "postulated" or "unidentified" component of the assessment.

Priorities within MMS now limit detached analysis of prospects to post-sale evaluation of tracts having received bids. Because the resource assessment process is directly related, and dependent upon, the lease sale process, more detailed mapping and evaluation of prospects as a routine treatment of tracts prior to sales could yield a more defensible result.

In the panel's view, MMS tends to emphasize the economic assessment of specific tracts for lease sale purposes. A policy of greater emphasis on resource base assessment process within MMS, could result in more definitive knowledge of the resource base available for long-term exploration and development under different political, technological and economic scenarios. This, in turn, could provide better support for enlightened decision making.

SOUTH FLORIDA BASIN OUTER CONTINENTAL SHELF, SOUTH OF 26° LATITUDE

In order to evaluate the data and methodology used in estimating oil and gas resources in the South Florida Basin OCS, the panel held a two-day meeting on June 27-28, 1989, at the regional offices of MMS in Metairie, Louisiana. The meeting was conducted in an informal, work-session format with excellent cooperation from the MMS regional supervisor for resource evaluation and his staff. Eight MMS professional staff members participated in the meeting.

Evaluations of Methodology

Geology/Geophysics Data Base

As their studies of the South Florida Basin were in an early stage of development and were ongoing, according to MMS staff, there is a limited data base. Extensive use of the available geologic literature was made in defining the nature of the basin and in focusing on the primary objective zones. No wells have been drilled in the offshore part of the basin to date, but well data from 6 dry holes along the Sarasota Arch, 11 dry hole along the Pine Key Arch, north and south of the basin, respectively, 3 dry holes in state waters on the south flank of the Sarasota Arch to the northeast of the basin, and data from the onshore, productive Sunniland Trend were extrapolated to define potentially productive zones. These data consisted of various types of well logs, sample logs and well histories. At this stage of the studies, no samples had been examined by MMS personnel, but the panel was informed that this is planned for the near future.

Seismic data consisted of commercial seismic lines of good quality ranging in age from 1981 through 1988. The highest density of the seismic grid pattern was concentrated on and around acreage blocks associated with prior lease sales. Several Lower Cretaceous horizons and top of basement were identified on the seismic lines and were used to construct structure contour maps. A limited amount of geochemical data is commercially available for most of the wells drilled along the southern flank of the basin, but this has not been included in the MMS assessment. Total organic carbon sufficient for generating hydrocarbons is present in the onshore Sunniland source rocks. Thermal history plots for the onshore wells suggest that mature source rocks may be present offshore.

Lower Cretaceous formations make up most of the prospective section. Prospective horizons are the Dollar Bay carbonate and evaporite sequence, the Sunniland Formation, which has produced over 92 million barrels of oil from onshore fields, the porous Brown dolomite, the Pumpkin Bay and deeper carbonate sequence and the Lower Cretaceous Shelf Edge Reef Trend. Additional potential exists in Triassic(?) basal sands over the basement surface and in grabens. The presence of porous, prospective sections in the wells surrounding the basin leads to the interpretation that there may be zones of comparable reservoir quality within the basin itself.

A total of 141 structural traps have been mapped, using the available seismic grid. The traps are predominantly anticlinal closures and fault closures. An additional 119 structures have been interpreted by projecting the number of mapped prospects from areas of good seismic control into areas where seismic coverage is sparse or nonexistent.

Resource Assessment Procedures

The absence of wells drilled in the offshore parts of the basin results in greater uncertainty in resource estimation. Information obtained from wells aids in interpreting the seismic data through refinement of velocity profiles for the stratigraphic section and in correlating seismic reflectors with specific subsurface rock units. Furthermore, logs, samples, and cores from wells permit the identification and evaluation of potential source and reservoir rocks. If successful, these wells can provide for direct detection of hydrocarbons. The lack of such analytical tools impedes the assessment process. Although the panel is not in a position to determine either the number or the location of wells needed to characterize adequately the hydrocarbon potential of the basin, at least one stratigraphic test well is necessary for the reasons cited above.

Because of the nonuniform distribution of seismic control, it was necessary to extrapolate from areas of dense seismic grid where traps were well-defined to areas of no control in order to estimate the total number of possible traps in the resource assessment area. This extrapolation was done by direct analogy: the number of and size distribution of unidentified prospects in areas of sparse control are chosen to match experience in areas with good seismic control. Most of the identified traps are structural (e.g., fault or fold closures), and the estimated total number of traps is considered to be conservative. The panel had some concern with the narrow range of uncertainty expressed about the number of unidentified prospects at the time of its meeting with this branch of MMS.

Lack of data within the South Florida Basin generally has prevented the identification of stratigraphic traps (e.g., pinchouts or unconformities) to the extent to which they are probably present. Stratigraphic traps are very common in the equivalent stratigraphic intervals throughout the Gulf Coast region, and it is reasonable to expect that they will become increasingly important in this basin as the data necessary to identify them develop.

Various estimates of trap filling were used in calculating volumetric estimates of the resources of the basin, but the calculations were not complete at the time of the meeting. In the interim, calculated ultimate recoveries from other Lower Cretaceous fields along the Gulf Coast trend were used as indirect analogs to provide an indication of what order of magnitude might be expected in the resource estimate.

The current method of assessing unidentified prospects based on a sample of seismically generated prospects is subjective in character and might be usefully supplemented in future assessments with objective statistical methods specifically tailored to a seismic search for prospects.

The procedure for assigning risks to prospects follows standard MMS practice: judgmental probabilities of .90 to .94 for the prospect risk in frontier provinces is reasonable in light of experience elsewhere. In this assessment, MMS did not encounter the problem of assigning multiple zone risks. However, as exploration proceeds and information accumulates, the need will emerge to appraise multiple zone risks systematically. The committee will address this subject in more detail in its report.

Conclusions

At this writing, MMS is in the process of determining its resource estimate for the South Florida Basin but does not have it completed. The MMS regional supervisor for the Gulf Coast confirmed on August 29, 1989, that the resource estimate was in Washington pending final review. As the resource estimate of oil and gas for the basin was not available from MMS at the time this report was completed, the panel was not able to assess its credibility.

Although the available data are limited to the South Florida Basin, it is apparent to the panel from its meeting with MMS staff that they are using appropriate techniques for interpreting the geological potential of the basin. The methods used in predicting the number, distribution, and volume of structural traps are appropriate for this frontier area at this stage in its history. Because the estimation procedure does not include prediction of stratigraphic traps, the panel believes that the total resource estimate will probably be very conservative.

The Committee on Undiscovered Oil and Gas Resources (the parent committee for this panel) will comment further on the resource estimate of the South Florida Basin in it report to be issued in March 1990.

Appendix B

Adequacy of the Data Base for Hydrocarbon Estimates of the Georges Bank Area of the North Atlantic Outer Continental Shelf

Committee on Undiscovered Oil and Gas Resources
Board on Earth Sciences and Resources
Commission on Physical Sciences, Mathematics, and Resources
National Research Council

National Academy Press
Washington, D.C. 1990

PREFACE

As part of its responsibility for managing the mineral resources of the U.S. coastal waters of the Outer Continental Shelf (OCS) that lie beyond the states' limits, the Minerals Management Service (MMS) of the U.S. Department of the Interior (DOI) leases areas on the OCS for exploration and exploitation of oil and gas resources. Since 1984, however, Congress has declared a moratorium annually on oil and gas exploration, development, and production in the Georges Bank area of the North Atlantic OCS because of environmental concerns. To address these concerns, the MMS asked the National Research Council (NRC) in 1988 to evaluate the adequacy of the data that the MMS uses to assess the environmental impacts of hydrocarbon exploration and production in the Georges Bank basin. This study was assigned to the Board on Environmental Studies and Toxicology. A subordinate issue—the adequacy of the information used by the MMS to assess the hydrocarbon resources potentially available for exploitation in the Georges Bank basin—was addressed by the Board on Earth Sciences and Resources (BESR) and is the subject of this report.

A few months before being asked to participate in the Georges Bank study, BESR had established a committee to study the methodology and data base used by the U.S. Geological Survey (USGS) and the MMS in their joint national assessment of undiscovered oil and gas resources (U.S. Geological Survey and Minerals Management Services, U.S. Department of the Interior. 1989. Estimates of Undiscovered Conventional Oil and Gas Resources in the United States—A Part of the Nation's Energy Endowment). Because of its particular expertise, this group, the Committee on Undiscovered Oil and Gas Resources, was assigned the subtask of evaluating the data base for the estimates of hydrocarbon resources in the Georges Bank basin made by MMS in conjunction with its lease sale activities. A panel of experts drawn from the full committee was appointed to do the review.

The committee accepted the additional task, which was to be carried out simultaneously with its larger study, with the understanding that its report on the Georges Bank basin would include only those considerations that the committee believed it could address without compromising the overlapping issues that were to be examined in the larger study. The Georges Bank report was scheduled for completion before the report on the national assessment, and the committee had not completed its evaluation of the quantitative methodology used in the national assessment when the Georges Bank report was written. Because the MMS uses the same statistical methodology and applies the same economic considerations for its regional resource evaluation as it did for the national assessment, the panel of the full committee that reviewed the Georges Bank data base was unable to address the issue of the quantitative statistical methodology in its report.

Besides reviewing the data base according to the charge to the committee, however, the report also addresses aspects of the MMS's approach to assessing the resource potential of the Georges Bank basin, such as how the MMS used the geological, geophysical, and geochemical data base, the validity of the MMS's extrapolations to areas lacking data, the MMS's practice of basing estimates of critical parameters on single numbers (point estimates) rather than on a range of values, and the adequacy of the MMS's considerations of uncertainty in its assessments. The report also assesses the impact of these practices on the overall estimates of hydrocarbon resources for the Georges Bank.

To examine the information available to the MMS and to evaluate the adequacy of the staff and facilities for making a credible estimate of the hydrocarbon potential of the Georges Bank area, the panel met at the MMS's North Atlantic Region office in Herndon, Virginia, with MMS staff members and supervisors responsible for the Georges Bank resource evaluation. Excellent cooperation and logistical support from Bruce Weetman, the MMS's regional director for the Atlantic Region, and Maher Ibrahim, MMS's regional supervisor for resource evaluation, assured the effectiveness of the meeting, which was conducted as an informal work session.

The geology and geologic history of the North Atlantic continental margin were reviewed, as were the processes by which hydrocarbons are generated, accumulated, and migrated. The panel then reviewed the nature, quality, and extent of the geological, geophysical, and geochemical data bases used in the resource estimation process. Panel members examined in detail representative samples of the data and reviewed the methods used to identify prospects and plays, as well as details of typical prospects within exploration plays. Finally, the panel examined the procedures by which the MMS determined the potential resources of the basin.

This report on the adequacy of the data base used by the MMS to estimate the hydrocarbon potential of the Georges Bank basin and on the credibility of that estimate is based mainly on the panel's meeting with the MMS's North Atlantic Region office staff, but it also draws on information and deliberations relating to the national assessment, because the MMS's methods and procedures have been essentially the same for both activities.

The Committee on Undiscovered Oil and Gas Resources expresses its appreciation to the MMS and its officials, whose cooperation and organized efforts resulted in a productive and informative meeting that led to the observations and evaluations described in this report.

CONTENTS

EXECUTIVE SUMMARY

In response to a request made to the National Research Council by the Minerals Management Service (MMS), a panel of the Committee on Undiscovered Oil and Gas Resources, operating under the Board on Earth Sciences and Resources, has reviewed the adequacy of the data base used by the MMS to assess the potential amount of undiscovered oil and gas present in the Georges Bank portion of the Outer Continental Shelf (OCS). The panel conducted a detailed examination of information available to the MMS and reviewed the adequacy of the data and the manner in which it was used to produce an assessment of the undiscovered hydrocarbon potential of the Georges Bank area.

The panel found several factors that make the estimates of oil and gas resources in the Georges Bank basin unduly conservative. Further, the estimates are presented in a way that tends to mask their uncertainty. This report therefore, recommends steps to improve the data available to the MMS and the methodology used in the analysis of those data.

In the Georges Bank portion of the OCS, the MMS has identified five exploration play types, i.e., potential oil or gas fields that are geologically related and have similar hydrocarbon sources, reservoirs, and traps. The only wells that have been drilled in the entire basin are in one play type in one part of the basin. While those wells—two Continental Offshore Stratigraphic Test (COST) wells and eight unsuccessful industry exploratory wells—have provided a large amount of stratigraphic, geochemical, and geophysical data, those data do not necessarily hold for the other four play types.

The MMS has a vast amount of seismic reflection data ranging in quality from poor to excellent and covering a wide area of the Georges Bank portion of the OCS. Because of budget limitations, however, the number of new additions, especially of reprocessed data, is often less than adequate.

Because information from wells is lacking for all but one play in the basin, the estimates of undiscovered hydrocarbon resources are based primarily on the identification of structures from the available geophysical information. As in any frontier province lacking established production, the initial efforts to locate hydrocarbon accumulations have been focused on the most obvious structural traps, which results in the less obvious stratigraphic and combination-type traps initially remaining unrecognized. This practice contributes to the geological uncertainty in the MMS evaluation because it overlooks the possible contribution of unidentified structural and stratigraphic trapping to the resource base. In addition, variations in porosity may go undetected, yet they may form stratigraphic traps with as much potential as areas of uniform porosity arched over low-relief structures.

Extrapolations of drilling results to the other plays in the basin should take into account the large degree of uncertainty involved in trying to estimate the resource base in the undrilled portion of the basin. The panel believes that a careful re-examination of the other plays in the basin using appropriate analogs from comparable sedimentary environments, especially those in Canadian waters to the north, could result in substantial modification of MMS estimates of the resource base with very little additional drilling. No attempt was made to quantify this judgment because it was beyond the scope of this study.

Although geochemical analyses based on data from the wells show that strata within the main sub-basin are organically lean, the type of kerogen in the older sedimentary rocks commonly produces gas, while the type of kerogen in the younger Cretaceous strata generally produces oil under appropriate thermal conditions. The hydrocarbon source in the COST G-1 well is reported by the MMS to be predominantly gas-producing, but the source in the COST G-2 well, carbonates and shales, is more oil-prone, with some gas associated with it.

The MMS's practice of assigning a single numerical estimate to the total number of prospects in a play, which includes both identified and unidentified prospects, masks the wide band of uncertainty that should be attached to this quantity. Also, because it is highly unlikely that all the prospects have been identified in any play type in this frontier province, the fact that the MMS's estimate of the total prospects is approximately equal to the number of identified prospects indicates that the MMS has underestimated severely the true number of prospects in the basin.

The lack of success in the one explored area of the Georges Bank basin appears to have led to low estimates of potential resources for the rest of the area. The MMS's assessment leads to an estimated mean of 40 million barrels of economically recoverable oil. If interpreted without taking into consideration the large degree of uncertainty resulting from the lack of testing of four of the five play types, the MMS's estimate implies an expectation of no more than one Class-B-size field of economically recoverable oil in the basin. Even though the basin seems to be primarily gas-prone, significant oil accumulations remain a good possibility based on geochemical analyses of the carbonates.

The weight given to the unsuccessful exploration results in the Baltimore Canyon area was not balanced by consideration of important discoveries in the Canadian areas of the OCS to the north of Georges Bank. This tended to result in underestimation of the resource potential.

Because of MMS's approach to the data, the absence of wells in four of the five play types, and the paucity of the latest new and reprocessed seismic data, it is the judgment of the panel that the MMS's assessment of undiscovered resources for the entire basin is very conservative.

Recommendations from the panel are in two categories: (1) the need for more data to support credible estimates and (2) steps to ensure careful reconsideration of the resource estimation.

To enhance the data base and to reduce uncertainty in the estimates of oil and gas resources in the Georges Bank basin, the panel recommends the following:

1. Drilling into each of the identified plays, for the purpose of obtaining geochemical and stratigraphic information for each play;
2. Increasing the MMS's staff access to reprocessed and more recent seismic data for all parts of the basin;
3. Upgrading to industry-standard, state-of-the-art data-processing equipment and techniques by the MMS for play and prospect evaluations.

For reconsideration of the resource estimate of the Georges Bank, the MMS should do the following:

1. Conduct a careful study of the Canadian basin discoveries and, using play summary maps, attempt to identify any favorable trends that can be projected into the U.S. portion of the Georges Bank.
2. Obtain reprocessed seismic data for portions of the reef trend to evaluate more fully the reservoir characteristics of these strata.
3. Provide a range of numbers of prospects in each play type to reflect more closely the uncertainty in the actual number of prospects that may exist.

4. Report unrisked assessment volumes along with the risk factors.

5. Reevaluate the lower limit for the range of acreage assigned to each prospect.

The Committee on Undiscovered Oil and Gas Resources concurred with the panel's findings and in its forthcoming report will elaborate further on the MMS's resource estimation methodologies and procedures, which, in the committee's judgment, tend to produce overly conservative estimates of undiscovered oil and gas resources.

1
INTRODUCTION

In 1988, the Minerals Management Service (MMS) requested the National Research Council (NRC) to undertake an examination of MMS hydrocarbon resource estimates of the Georges Bank area of the North Atlantic Outer Continental Shelf (OCS). This study was assigned to the Board on Mineral and Energy Resources (BMER) [now the Board on Earth Sciences and Resources (BESR)]. At the time of this request, the board had already initiated an examination of the methodologies and procedures used by the MMS and the U.S. Geological Survey (USGS) in their most recent joint national assessment of undiscovered conventional oil and gas resources of the United States (U.S. Department of the Interior, 1989). The Committee on Undiscovered Oil and Gas Resources that had been established to conduct the larger review of MMS and USGS methodologies and procedures was assigned the examination of the Georges Bank resource assessment.

The specific charge to the committee with respect to the Georges Bank area was as follows:

> Assessments will be conducted of the adequacy of the scientific and technical information to evaluate the estimates of the hydrocarbon resource base of the designated moratorium area of the North Atlantic OCS. The assessments will include an examination of
>
> • what is known, plus reasonable extrapolations, accompanied by an expression of the error or uncertainty;
> • what is not known, and the reasons why (e.g., difficulty of measurement, confounding of data, lack of theory, lack of data, insufficient time); and

• what could be known with various reasonable additional investigative resources (manpower, money, facilities) and time.

The above assessments will be made in the context of the MMS's need to support leasing decisions and to predict and manage environmental effects of OCS oil and gas activities.

The committee set up a panel on Georges Bank from among its members to address this task. A related study of environmental impacts was done simultaneously by the NRC's Board on Environmental Studies and Toxicology (BEST).

The panel met with the MMS staff at their office in Herndon, Virginia, where the data used for the Georges Bank estimate and for the entire U.S. North Atlantic OCS are maintained. The geology of the Georges Bank was reviewed, and the geological and geophysical data used in the resource estimation process were examined to evaluate the nature, quality, and extent of the data base. The geochemical characteristics of the basin were reviewed, and information related to the gas- or oil-prone nature of the sedimentary strata was examined. The methods used in identifying prospects[1] and establishing exploration plays[2] were examined and typical prospects within exploration plays were reviewed.

After reviewing information relating to the factors that cause hydrocarbon accumulations, the panel examined the procedures by which the potential resources of the basin were determined and made a judgment as to the adequacy of the assessment of undiscovered oil and gas resources assigned by the MMS to the Georges Bank basin.

Although the committee concurred with the panel's findings, as presented herein they are to be considered preliminary and are subject to review and possible modification by the committee in its forthcoming report on the joint national assessment conducted by the MMS and USGS.

---

[1]Prospect — A potential oil or gas field.

[2]Play - A group of geologically related prospects with similar hydrocarbon sources, reservoirs and traps.

Location

Georges Bank is the most southwestern of a series of banks on the North Atlantic Outer Continental Shelf (OCS) that parallel the coast between Newfoundland and Nantucket Island (Fig. 1). It is located in a region of numerous deep submarine canyons, but the bank has a moderate water depth commonly less than 260 feet. The Georges Bank is a trough-shaped, extensional basin, about 200 miles long by 80 miles wide. On average it contains in excess of 26,000 feet in thickness of Mesozoic and Cenozoic sedimentary strata, with several deeper sub-basins containing as much as 40,000 feet in thickness of sedimentary rocks.

A dispute regarding the boundary between the Canadian and U.S. portions of the basin was resolved by the International Court of Justice in a ruling that defined the location of the boundary. The approximate location of that boundary is shown in Figure 1.

The Georges Bank basin is the northernmost of seven geologically defined provinces constituting the MMS subdivisions of the Atlantic OCS (Region 9-A). It is also the northernmost of the four MMS planning areas of the Atlantic OCS for lease sale and regulatory purposes.

Geologic History

The Georges Bank area provides evidence for a generally accepted geologic history that includes the entire North Atlantic region. Older sedimentary strata, accumulated prior to the rifting and separation of a large continental mass into the present-day continents, lie below younger marine and terrigenous deposits formed in the Triassic and Jurassic Periods during and after continental rifting. The rock types seem to have been controlled mainly by ancient seafloor topography and the distance from ancient shorelines. The sedimentary rocks deposited before and during the rifting stages are now preserved in rift grabens and half-grabens; they are the reworked products of the erosion of a thick section of Paleozoic and Triassic clastic, volcaniclastic, and fresh water deposits. These grabens formed in the crystalline and metasedimentary basement rocks.

The graben-filling deposits were further faulted and deformed during the long rifting process. During the Triassic, as rifting began, the region was uplifted and then reduced to an erosional plane, forming an unconformity (the breakup unconformity) above which Triassic coastal sedimentary rocks were deposited.

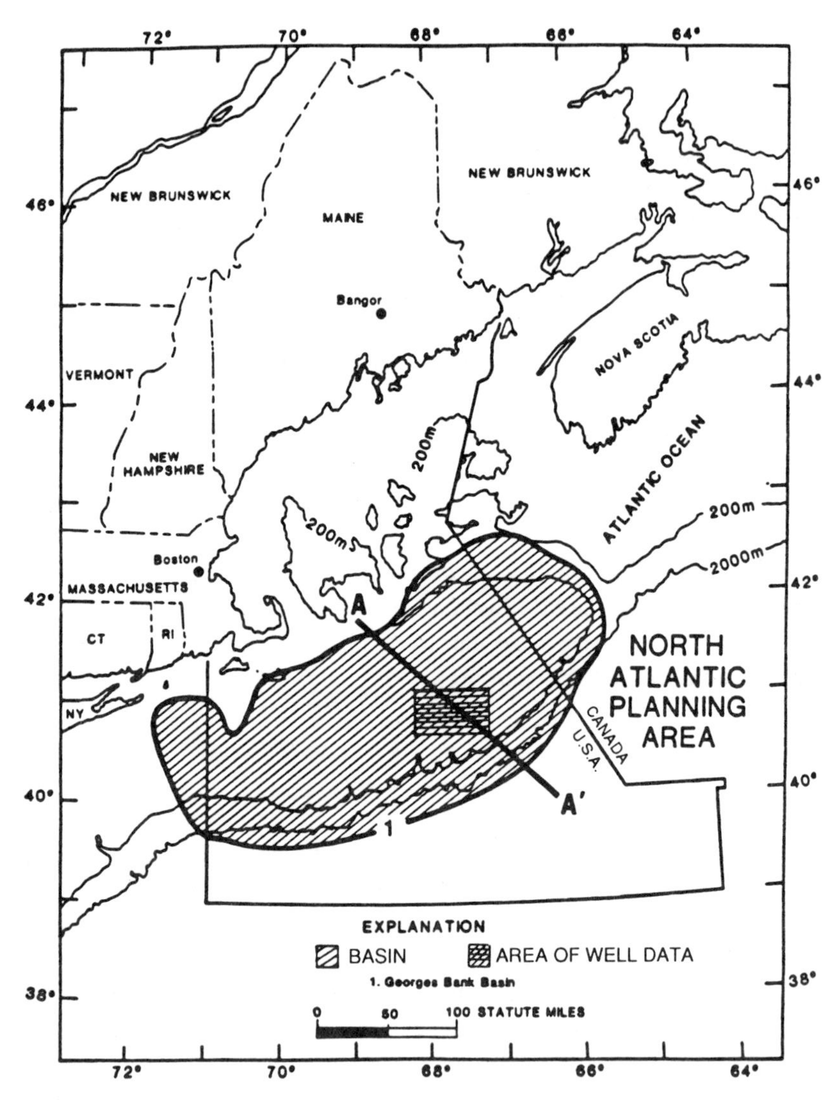

FIGURE 1 Index map of the Georges Bank area showing location of COST and exploratory wells (rectangular grid), and location of cross section (A-A′) shown in Figure 2. SOURCE: Minerals Management Service, Atlantic Region.

Distinctive episodes of rifting continued into Jurassic time. After the initial rifting, marine waters inundated the area during Late Triassic time, depositing shallow marine and shoreline sediments. These included limestones, dolomites, and evaporites (mostly salt).

Shoreward, this sequence contains clastic rocks and coal beds. Seaward, large thicknesses of evaporites were deposited. The evaporites were subsequently deformed by sediment loading and produced a number of salt diapirs in the northeastern part of the area. A widespread erosional unconformity caps the Triassic coastal, shallow marine sequences.

An Early Jurassic rifting episode renewed subsidence, resulting in the widespread deposition of a sandstone that was subsequently subaerially exposed. Rift-related tectonics ended by Middle Jurassic time, when Atlantic basin subsidence and changes in sea level caused more open marine conditions and the consequent deposition of limestones (some of which were later altered to dolomite). The lowering of sea level, shortly after the beginning of the Late Jurassic, produced a pronounced eastward (seaward) shift in the limit of clastic sedimentation. Except for periodic fluctuations of relatively short duration, sea level has remained near its present position since Late Jurassic time. Therefore, Cretaceous and Tertiary sedimentary strata are nearly all sandstones, shales, and siltstones with some thin, interbedded limestones.

A pronounced shelf edge was established by the middle of Late Jurassic time as the ocean basin continued to subside (Fig. 2). Carbonate detritus, consisting of shells, coral fragments, and other biogenic products, accumulated along the outer shelf edge, forming a carbonate reef that is distributed discontinuously and is parallel with the present shelf break. Carbonate accumulation persisted into the Early Cretaceous, when it was terminated, probably because of climate and sea-level changes.

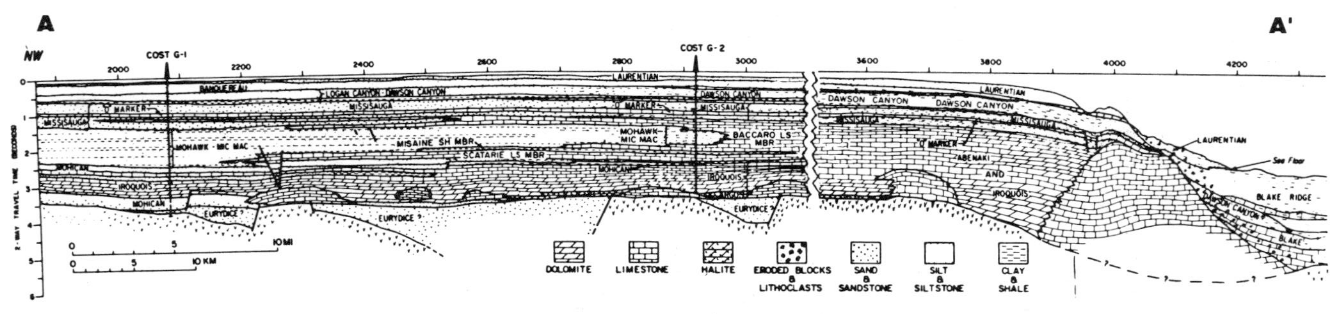

FIGURE 2 Cross Section A-A′ showing stratigraphic section in the two COST wells and across the Georges Bank basin interpreted from seismic profile. Location given in Figure 1. V symbols at base of sedimentary strata indicate acoustical basement. SOURCE: Modified from Poag, C. W., 1982.

2
GEOLOGICAL AND GEOPHYSICAL DATA BASE

During the period 1966 to 1988, 111,000 line miles of Common Depth Point (CDP) seismic data were collected in industry-supported offshore seismic surveys in the Georges Bank (Fig. 3). Of this total, the MMS North Atlantic regional office has in its possession data on approximately 62,000 line miles in the U.S. portion of Georges Bank and 6,000 line miles in the Canadian portion. In addition, the MMS also has a substantial quantity of time-migrated seismic data, depth-migrated seismic data, relative-amplitude data, velocity scans, and gravity and magnetic data.

The seismic grid, ranging from 1/8 mile by 1/8 mile in the lease sale areas to 3 miles by 3 miles outside, as well as other geophysical parameters, were chosen by the industry using standard, state-of-the-art geophysical practice adequate for locating potential oil-bearing structures. The seismic grid for the Georges Bank basin is spaced sufficiently to identify significant structures that may contain hydrocarbons, but the processing of some lines may not be adequate to identify all of the existing structures in a particular play.

As in any large geophysical survey, the quality of data characterizing the Georges Bank basin ranges from fair to very good, with the poorest-quality data having been acquired before 1975. Also, as would be expected, the conventional CDP data obtained over the deep-water canyon areas are of poorer quality. However, to correct this inadequacy, the MMS obtained specially processed data (by wave-equation depth migration) from the Exxon Corporation to gain a better picture of the structures in the deep-water canyon areas.

The MMS systematically updates its geophysical data base with new data, as well as reprocessed data, as soon as they become available. Because of budget limitations, however, the number of new additions, especially of reprocessed

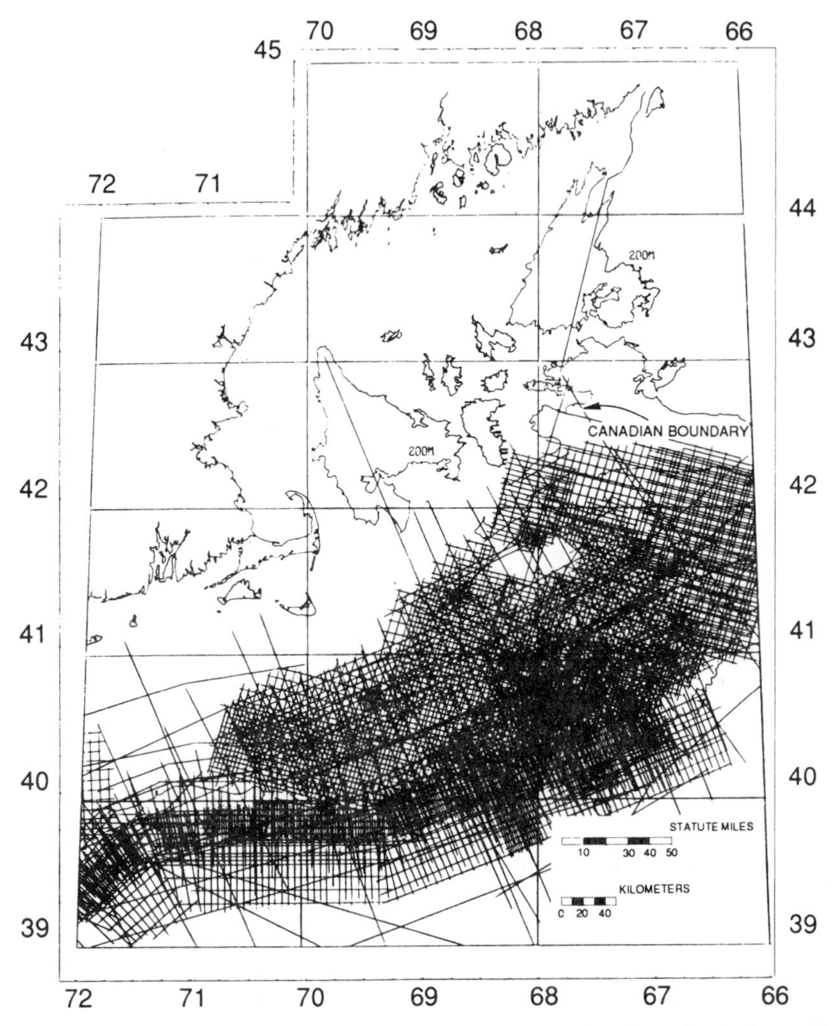

FIGURE 3 Index map of the Georges Bank basin area showing lines of seismic profiles available to the Minerals Management Service. High concentration of seismic lines in vicinity of 41°N latitude/68°W longitude is location of two COST wells and several industry exploratory wells. SOURCE: Minerals Management Service, Atlantic Region.

data, is often less than adequate. The highly qualified geophysical staff of the MMS has been able to analyze and interpret this massive collection of geophysical data using industry-accepted methods and standard techniques despite their small number.

The seismic data were integrated with the available geologic data taken from a small number of wells in an effort to relate seismic reflection events to specific geologic horizons. Only ten wells have been drilled in a limited area of the basin (see Fig. 1). The first two were the Continental Offshore Stratigraphic Test (COST) G-1 and G-2 wells, which were drilled by a consortium of companies to establish the geological, geophysical, and geochemical characteristics of the basin. The objective of both COST wells was to provide as much stratigraphic information as possible about the basin—not to discover oil or gas. (As a matter of MMS policy, COST wells are located away from structures to avoid potential accumulations of hydrocarbons.) Subsequently, eight exploratory wildcat wells were drilled by Exxon, Conoco, Tenneco, Mobil, and Shell; all eight proved to be dry holes, except for some shows of wet gas.

The MMS obtained all pertinent information on the COST wells. These data include engineering and operations reports, well logs, well surveys, seismic check-shot surveys, core descriptions, core photographs, petrophysical analyses of core plugs, washed and dried cutting samples, and canned samples for geochemical analyses. In addition, a full range of information was available to the MMS on a number of industry wells that were drilled after the COST wells and were subjected to similar geoscientific evaluation. MMS geoscientists conducted lithological and petrographic analyses, paleontologic and depositional environmental studies, log analyses and formation evaluations, and optical kerogen and molecular geochemical analyses on the available well data. From the data, they produced play and prospect maps.

The two COST wells (G-1 and G-2) served as keys for determining for that play (1) the nature of the stratigraphic sections (lithofacies, reservoir facies, and so on) penetrated in this area, (2) the levels of geothermal (time-temperature) diagenesis of the sedimentary strata throughout geologic time, (3) the nature of the resident organic matter characterizing each of the differing lithofacies, and (4) the organic carbon content and hydrocarbon richness of the various strata encountered. These data provided the foundation of a geochemical-geological framework for this area. Complementary (although not as complete) geochemical information was available to the MMS from the eight subsequently drilled industry wells. In addition, the MMS has acquired geochemical data from surveys and bottom samples aimed at direct detection of hydrocarbons.

The MMS makes some of the data available to the public by including information from each COST well in published OCS reports. In addition to these reports, the MMS has published a report on one of the industry-drilled wells, the Exxon Lydonia Canyon 133-1 well.

All of the Georges Bank industry and COST wells are clustered in the main sub-basin of the province, representing investigation of about 5 percent of the basin area and about 3 percent of the volume of sedimentary strata in the basin (see Fig. 1). Most of the basin is devoid of well drilling and the data such drilling provides. While this situation is not unusual in the exploration of new oil and gas provinces, the absence of wells in a large part of the basin requires that most of the judgments pertaining to the resource potential and decisions concerning the potential of undrilled prospects outside the sub-basin be based solely on geophysical data and comparison with analogous geological areas that have been more extensively drilled.

<u>Although the MMS has done a professionally competent job in interpreting the available geophysical data, it must be noted that the use of geophysical data alone for estimating oil and gas resources has limitations. Any procedure that relies solely on such information is beset by uncertainties that affect estimates of the undiscovered oil and gas resources estimates of the area.</u>

In comparing analogous geological areas, the MMS relied primarily on the disappointing exploration experience in the Baltimore Canyon area to the south of the Georges Bank basin. No consideration was given to the more successful exploration results in the Canadian basins to the north.

3
PETROLEUM GEOLOGY

The MMS has identified five general exploration play types[3] in the Georges Bank basin: (1) Triassic rift basins filled with Triassic and Lower Jurassic sedimentary strata, (2) anticlines on the continental shelf (Jurassic and Lower Cretaceous strata), (3) shelf edge, carbonate reef trend with discontinuous areas of enhanced porosity in Upper Jurassic and Lower Cretaceous rocks, (4) Upper Jurassic and Lower Cretaceous sedimentary wedgeouts against the slope and seaward edge of the reef trend, and (5) structural traps seaward of the continental shelf in Upper Jurassic and Lower Cretaceous strata. The Triassic rift-basin play has been subdivided into three categories of trap types: (a) structures within the Triassic rift basins, (b) stratigraphic traps in Triassic rift basins, and (c) stratigraphic traps in rift basins formed by only one fault downthrown in the landward direction (half-grabens). Part of the rift-basin play underlies the continental shelf anticline play.

The ten wells (two COST wells and eight industry wells), localized in a small area, evaluated a single play type—Jurassic anticlines on the continental shelf. The COST wells tested two of the sedimentary sections of interest in this area, an inner neritic-type clastic sequence penetrated by the COST G-1 well, and a thick shelf and reef carbonate sequence penetrated by the COST G-2 well. However, both of these classes of sedimentary strata relate only to the shelf anticline play. The wildcat wells, as mentioned previously, were unsuccessful, containing only shows of wet gas.

Seismic data, confirmed by well information in the limited area where well control is available, indicate that the anticlinal structures tend to be less

---

[3] Play type — The specific geologic structure or rock type that characterizes a play.

pronounced at shallow depths. In part, sediments draped over basement uplifts have blanketed and modified the well-developed deeper structures.

It should be emphasized that, in areas where structures are not large or well developed, local variations in porosity both areally and at depth may form stratigraphic traps with as much potential as areas of uniform porosity arched over low-relief structures. The relative timing of the formation of traps, predating the generation and migration of hydrocarbons from the source rocks to the traps, allows for possible accumulations. In addition, the occurrence of shales and anhydrites interbedded with potential reservoir rocks such as sandstone and carbonate seems to be adequate to seal any trapped accumulations.

While the MMS has mapped three seismic horizons in the Georges Bank basin, only one of the mapped horizons (approximately Middle Jurassic) has been used, and then only as an average structural indicator in the resource assessment. Each play has been defined on a summary sheet that includes the play description and parameters that characterize the play, such as porosity, thickness, acreage, and so on. In addition, a trap file is maintained for individually identified prospects to document the essential elements of each. Analogs of fields producing from similar traps have also been identified to establish a possible order of magnitude of oil and gas accumulation that might be expected if the prospect were found to be productive.

The MMS has defined the geological parameters that could produce hydrocarbon accumulations in this frontier province. As with any frontier province lacking established production, the initial efforts to locate hydrocarbon accumulations have been focused on the most obvious structural traps, which has resulted in the less obvious stratigraphic and combination-type traps initially remaining unrecognized. It is commonly accepted that once commercial production is established within a basin, a deliberate and aggressive search for additional traps, spurred by economic incentive, should produce a significantly larger number of prospects.

Geochemical analyses of samples from the COST G-1 and G-2 wells and from available industry wells indicate that the sedimentary strata within the main sub-basin of Georges Bank, where all the wells are situated, tend to be organically lean. Organic content is higher in shallow horizons (0.7 to 1.0% total organic content [TOC]) and is more dispersed with depth. Cretaceous strata seem to be more oil-prone, whereas the older sedimentary rocks contain kerogen types that, at maturation, commonly produce gas. Geothermal gradients indicate that oil should be generated above 12,000 feet and gas generated below that level.

The hydrocarbon source for the sandstones in the COST G-1 well, Jurassic and Cretaceous shales, is reported by the MMS to be predominantly gas-prone, and most of the analyzed samples show the organic content to be immature and lean. Rocks that occur in the middle zone (Zone B) in the well are immature but have the potential to be good sources for oil and associated gas if found in areas of higher thermal maturity (G. Bayliss, personal communication). In contrast, the source for the carbonate strata in the COST G-2 well, Jurassic carbonates and intercalated shales containing amorphous-sapropellic kerogen, is more oil-prone, with associated gas, especially in the lower part of the section. The organic content of the strata in Zone D (13,700 to 16,900 feet) are rated as fair to good sources of oil and associated gas at initial stages of petroleum liquid generation. The organic content of the carbonate sections in this well indicates that such strata will be of major significance to the hydrocarbon potential of the basin (G. Bayliss, personal communication). Moreover, these considerations with different types of organics at different stratigraphic levels, indicate more than one source for the hydrocarbons in a single play.

In assessing the hydrocarbon potential of carbonate-related oil plays, the importance of the dominant amorphous-sapropellic kerogen (oil-prone) facies in estimating hydrocarbon potential transcends both the low TOC content and the regionally lower thermal maturity characteristic of the area, because this type of kerogen begins producing oil liquids at lower levels of thermal maturity than all other types (Bayliss, 1985, Fig. 2). <u>Thus the limestones of the explored part of the Georges Bank with low organic carbon contents that are only moderately thermally mature can be very prolific sources for oil liquids because of their ideal oil-prone amorphous-sapropellic kerogen content. This situation suggests a greater potential for oil accumulations in the Jurassic anticline play than the MMS expects.</u>

Although the available geochemical data enable a reasonable assessment of the oil- and gas-generating potential of one sub-basin in the Georges Bank area, extrapolating this information to estimate such potential throughout the basin involves a substantial degree of uncertainty because of the absence of a complete network of wells. This deficiency is particularly relevant to the resource assessment, because prospects seaward of the reef edge are likely to exhibit an organic content much higher than that observed on the shelf and should become more sapropellic. The oil- and gas-generating potential seaward of the reef edge, and consequently the probable hydrocarbon resources, should be greater than the resources on the shelf because of the presence of more favorable organic source material (Bayliss, 1985, Figs. 2 and 3).

As previously mentioned, the two COST wells tested only two of the sedimentary sections of interest in this area—a thick, inner neritic-type clastic sequence (G-1 well), and a thick shelf and reef carbonate sequence (G-2 well). Reservoir-quality rocks are commonly present in the sandstones, whereas the development of porosity in the carbonate strata is more variable and facies-dependent. A trend of carbonate porosity is present along the reef, where it provides one of the trap types recognized by the MMS. In general, depending on the sequence of rocks encountered in a well, porosity tends to be greater at the shallower depths. As the source rocks associated with the carbonates of the area are more oil-prone than are the deeper clastic rocks, the carbonate reef trend should have a greater potential for liquids than for gas.

4
PETROLEUM RESOURCE ASSESSMENT OF THE GEORGES BANK

The MMS's summary of the undiscovered resource assessment of the Georges Bank basin is shown in Table 1.

TABLE 1. Probability Estimates of Undiscovered Oil and Gas in Georges Bank Basin (expressed in terms of volume versus percent chance).

	Crude Oil (billion bls)*			Natural Gas (tcf)**		
	95%	Mean	5%	95%	Mean	5%
Technically recoverable	0.00†	0.10	0.38	0.00	1.94	6.75
Economically recoverable	0.00	0.04	0.19	0.00	0.98	4.16

* Barrels
** Trillion cubic feet
† Reported as 0.00 because the probability of no technically (economically) recoverable oil (gas) is greater than 0.05.

SOURCE: U.S. Geological Survey and the Minerals Management Service, 1989.

Approximately 79 percent of the risked[4], economically recoverable resources reported by the MMS are assigned to two play types: anticlines on the continental shelf and structures within Triassic rift basins. The former play type

[4]Risk is the probability that no commercial oil or gas field exists.

is assigned about 48 percent and the latter play type about 31 percent of the total hydrocarbon resources of the basin. Accessibility is one factor in assigning a large proportion of the resource to these two plays types because they are superimposed and are located on the continental shelf where the relatively shallow water makes drilling easier and less expensive. Both of these play types are logical and primary targets in the early phases of exploration, but if they alone are emphasized, as the MMS has done, the possibility of large oil and gas accumulations in other play types—such as the shelf-edge carbonate reef trend, seaward sedimentary wedgeouts against the slope and seaward edge of the reef trend, and structural traps seaward of the continental shelf—may not be given proper consideration. Only 21 percent of the hydrocarbon resources in the Georges Bank basin are expected to come from the five other play types, even though prospects within these plays include carbonate facies that geochemical analysis indicates are more oil-prone than the sandstone facies.

As reported in the MMS's summary table (Table 1), a mean of only 40 million barrels of economically recoverable oil is estimated for the Georges Bank basin. This represents the equivalent produced by just one Class-B-size field. If interpreted without taking into consideration the very large degree of uncertainty that exists because of the lack of testing of four of the five play types and the limited testing in only one part of the basin of the other play type, the MMS's estimate implies an expectation of no more than one Class B economic oil field in the basin. Estimates of recoverable gas are slightly higher; 0.98 trillion cubic feet (tcf) of economically recoverable gas (Table 1) equals the output of about one national-class, giant gas field. Estimates such as those shown in Table 1 that do not reflect the large degree of uncertainty brought about by lack of data in the basins being evaluated lead to very conservative numbers, in the panel's judgment.

Data obtained from only the two COST wells and eight unsuccessful industry exploratory wells in one play, localized in the same sub-basin, should not have been extrapolated to characterize the average amount of economically recoverable undiscovered hydrocarbons in the entire basin, because the other plays are fundamentally different and must have independent hydrocarbon sources based on their structural and geographic locations.

The MMS assigns a numerical estimate for the number of prospects in each play type for input into the PRESTO[5] model. In several plays, the total number of prospects within the play was set equal to the number of identified prospects; that is, it was assumed that there are no unidentified prospects. In other cases, a range of numbers was entered when MMS geologists assumed that not all of the prospects in a play have been mapped. The assignment of a single number for the prospects in each play type, for whatever reason, does not show an adequate appreciation of the uncertainty in the current state of knowledge in this area.

It is highly unlikely that all the prospects have been identified in any play type in this frontier province. The seismic data and the fact that knowledge of the basin is very incomplete suggest that many unidentified prospects remain. Therefore a single numerical estimate of the total number of prospects in a play that is close to the number of identified prospects almost always underestimates the true number of prospects in the basin. In addition, the single value masks the wide band of uncertainty that should be attached to this quantity. Explicit representation of the current state of uncertainty about the number of prospects in the Georges Bank basin would indicate a degree of uncertainty about recoverable resources that is larger than that presented by the MMS.

The panel considers the range of acreage assigned to prospects to be too wide. For example, the acreage assigned for the shelf anticline play ranges from 6 to 36,780 acres for 90 prospects; the area assigned for the anticline play seaward of the reef trend ranges from 18 to 12,590 acres; the Triassic rift basin play is assigned a range from 2 to 16,583 acres. The lower limit of these ranges is so low as to be unreasonable.

The exploration history of the Baltimore Canyon about 500 km to the southwest of Georges Bank, which has produced poor results, appears to have played a major role in the MMS's resource evaluation of Georges Bank. In contrast, no consideration was given to the more successful exploration programs in the geologically similar Canadian areas to the north. The appropriate weight that should be given to these related areas can be determined by the use of play summary maps of the source reservoir and trap conditions in the Canadian sector to see if any favorable trends can be projected into the U.S. part of the area. Balanced consideration of the exploration history of geologically similar

---

[5]The acronym for Probabilistic Resource Estimates Offshore, a computer model developed by the MMS for the assessment of undiscovered hydrocarbons using a simulated exploration methodology.

contiguous areas would yield more realistic expectations than those of the current estimate.

In summary, the combined effects of several practices in the MMS's approach to resource estimation result in very conservative estimates of the potential oil and gas resources in the Georges Bank basin of the North Atlantic OCS.

5
CONCLUSIONS

The Georges Bank basin contains five identified play types, only one of which has been tested through drilling. That play is located in one small area of the basin. Most of the basin remains unexplored.

Although the seismic grid is adequate to identify significant structures present in the basin, the small amount of reprocessed seismic data the MMS is able to obtain limits the staff's ability to evaluate fully the potential hydrocarbon accumulations in some of the plays, especially those related to stratigraphic traps and to carbonate rocks that are in deeper water.

The estimates produced by the MMS for oil and gas resources in the Georges Bank basin are judged by the panel to be overly conservative because of the following MMS practices in its resource assessments:

1. Using single-number estimates instead of ranges of key quantities;

2. Assuming that the total number of prospects in a play is equal to the number of identified prospects;

3. Using only unsuccessful exploration results in analogous areas such as the Baltimore Canyon;

4. Extrapolating from areas of low potential, such as shelf anticline plays, to more promising areas, such as the carbonate reef or the area seaward of the reef, which have more favorable organic source material;

5. Ignoring the potential of stratigraphic traps in favor of the more obvious structural traps; and

6. Inadequately incorporating expressions of uncertainty in the estimation process, a necessary step given the lack of drill information for four of the five play types in the basin.

6
RECOMMENDATIONS

Recommendations fall into two categories: (1) the need for more data to support credible estimates and (2) steps to ensure careful reconsideration of the resource estimation.

Data Needs

With additional investigative resources, much needed information essential to a credible estimate of the hydrocarbon resources could be obtained that would reduce the uncertainty and lead to more realistic estimates. Such investigative resources include:

1. Drilling into each of the identified plays, for the purpose of obtaining geochemical and stratigraphic information for each play;

2. Increasing the MMS's staff access to reprocessed and more recent seismic data for all parts of the basin;

3. Upgrading to industry-standard, state-of-the-art data-processing equipment and techniques by the MMS for play and prospect evaluations.

Reconsideration of the Resource Evaluation

Much of the foregoing discussion suggests that to ensure a thorough reconsideration of its evaluation of the resource potential of the Georges Bank basin, the MMS should do the following:

1. Conduct a careful assessment of information from the exploration and related resource assessment activities for the Canadian basins to the north of Georges Bank. With that information, play summary maps should be constructed of the source, reservoir, and trap conditions of the Canadian discoveries to see if any favorable trends can be projected to the U.S. portion of the area. This would enhance greatly the MMS's ability to assess more effectively the undrilled portions of Georges Bank.

2. Obtain reprocessed seismic data for portions of the carbonate reef trend to evaluate more fully the reservoir characteristics of these strata.

3. Provide a range of numbers of prospects in each play type for input to the PRESTO model to reflect more accurately the uncertainty in the actual number of prospects that may exist in each play type.

4. Report unrisked estimated volumes of oil or gas in the prospects along with the risk factors.

5. Reevaluate the lower limit for the range of acreage assigned to each prospect.

A full discussion of the methodologies and procedures by which resource estimates made by the MMS tend to be systematically conservative is beyond the scope of this review of the data base, but will be addressed by the full committee in its report scheduled to be released in late 1990.

REFERENCES

Amato, Roger V. and Bebout, John W., eds. 1980. Geologic and Operational Summary, COST No. G-1 Well. Georges Bank Area, North Atlantic OCS. U.S. Geological Survey. Open-File Report 80-268.

Amato, Roger V. and Simonis, Edvardas K., eds. 1980. Geologic and Operational Summary, COST No. G-2 Well. Georges Bank Area, North Atlantic OCS. U.S. Geological Survey. Open-File Report 80-269.

Bayliss, G. 1985. Geochemistry in Oil and Gas Exploration: A Review. World Oil, August 1.

Poag, C. W. 1982. Stratigraphic Reference Section for Georges Bank Basin: Depositional Model for New England Passive Margin. Bulletin, American Association of Petroleum Geologists, vol. 66. no. 8, pp. 1020-1041.

Appendix C
The Effects of Play Mixing in the Permian Basin

In the Permian Basin, the USGS defined 10 plays made up of from 1 to 10 clusters from the Significant Oil and Gas Fields of the United States Data Base (see Table C-1). In plays 2, 3, 5, and 6, the committee found notable mixing of sandstone and carbonate lithologies. In plays 2 and 3, for example, 17 percent of the fields had sandstone lithologies, while 24 percent had carbonate lithologies. Play 5 (Delaware/Val Verde Basin Gas Play) contained 30 percent carbonate fields and the balance were sandstone fields. Play 6 (Eastern Shelf and Midland Basin Shelf Sequence) contained 31 percent sandstone fields, with the rest carbonate. As the differences in lithology suggest, these plays also contained significant mixing of different types of depositional systems. This mixing of lithologies and depositional systems appears to violate the common definition of "play."

The following discussion compares the discovery histories as they appear when analyzed according to the USGS's grouping of fields into plays with the discovery histories as they appear when fields grouped into plays have more consistent lithologies and depositional systems. The committee did not review the makeup of the clusters in the data base, nor was it charged to do so. This analysis presents a qualitative, non-statistical review of the use of clusters to formulate plays that suggests the need for more rigorous sensitivity studies of play formulation. The pattern established by the discovery histories varies between the plays as the USGS defined them and the plays as redefined for this analysis; it is very likely that the assessment of resources would also change with a different play grouping.

TABLE C-1. NRG Clusters Included in Permian Basin Plays

Play Number	Clusters Included in Play
1	301, 303, 313
2	305, 307, 309, 323, 333
	335, 337, 339, 357, 359
3	311, 317, 319, 321, 341
	347, 365
4	343
5	345, 355, 361, 363, 373
6	315, 325, 331, 349, 369
	371, 375, 377
7	367
8	379, 381, 383, 399, 401
9	385, 387, 395
10	389, 391, 393

PLAY 5

Analysis of Clusters 361 and 363

In play 5, clusters 355 and 361 contain major carbonate reservoirs, while clusters 363 and 373 contain major sandstone reservoirs. Clusters 361 and 363 represent different depositional systems: cluster 361 represents patch-reef to open-platform carbonate deposition, whereas cluster 363 represents fluvial-deltaic to barrier-strandplain sandstones.

For this analysis, the field discovery histories for clusters 361 and 363 were partitioned into thirds using the same dates as those used by the USGS (the last third of fields discovered was also divided into halves to give four dates). Field-size class versus number of fields in each date bin were then plotted as histograms for each date and also as a cumulative total. This procedure is the same for the several analyses that follow.

Comparing histograms shows some notable differences among them, especially in the cumulative plot for all four dates for each cluster (see Figures C.1 to C.3). Cluster 363 is skewed toward smaller field sizes; it has a maximum number in class 6 and no class 11 fields, while cluster 361 has a maximum

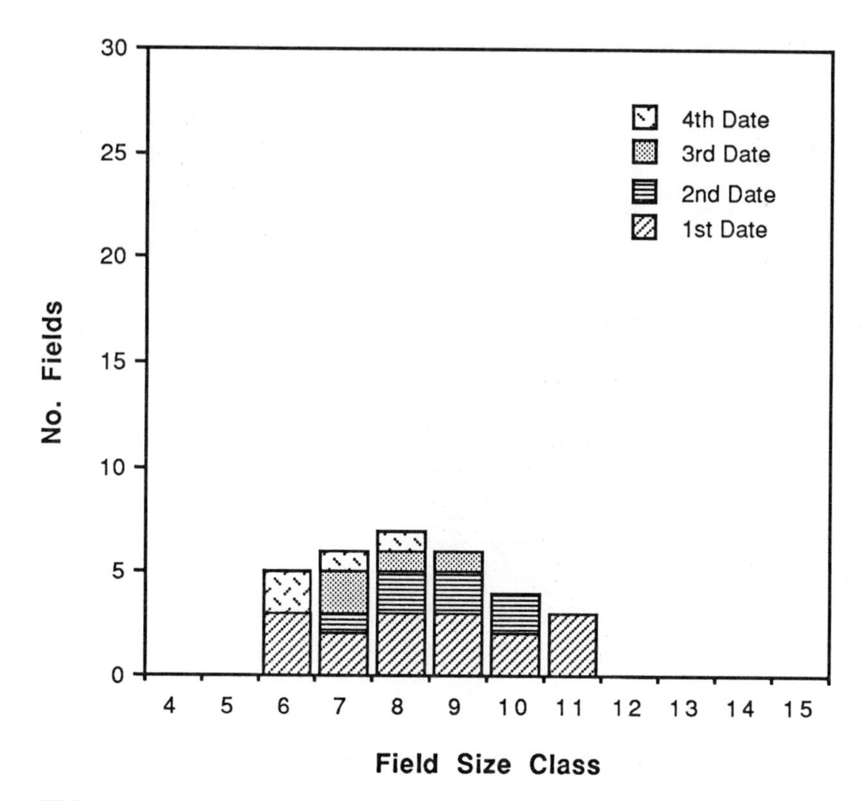

FIGURE C.1 Field-size distribution by discovery date for cluster 361. This cluster represents patch-reef to open-platform-carbonate deposition. Note that the histogram has a maximum frequency in class 8 and three fields in class 11.

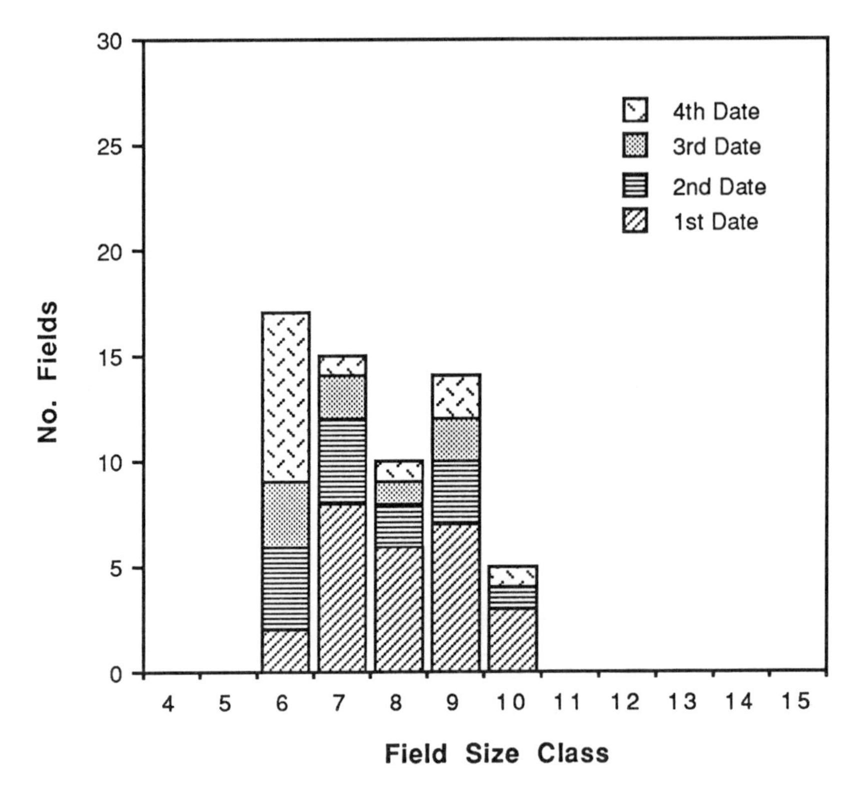

FIGURE C.2 Field-size distribution by discovery date for cluster 363. This cluster represents fluvial-deltaic to barrier-strandplain sandstones. Compared to the histogram for cluster 361, this histogram is skewed toward smaller field sizes.

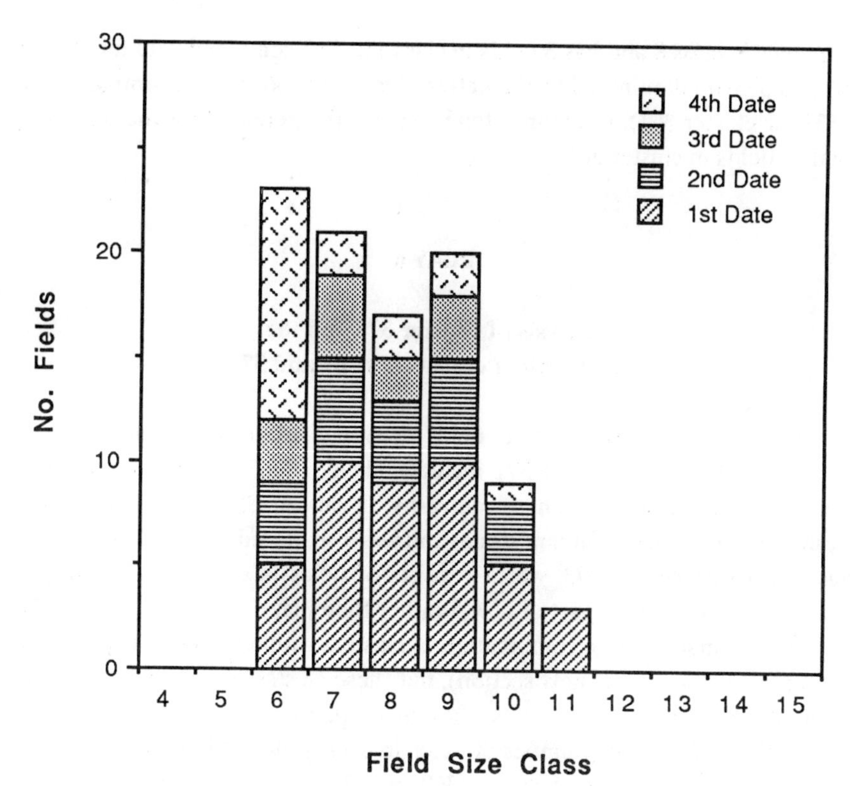

FIGURE C.3 Combined histogram for clusters 361 and 363. The pattern for cluster 363 dominates the combined data.

frequency in class 8 and has 3 fields in class 11. The combined data for clusters 361 and 363 are dominated by the pattern for cluster 363 (n=61 compared with n=31 for cluster 361), which may tend to mask the potential for the remaining smaller fields in cluster 361.

PLAY 6

Analysis of Cluster 375 and the Combination of Clusters 369 and 371

In play 6, cluster 369 includes fields developed in carbonate-open-platform to patch-reef reservoirs, and 371 includes shelf- edge carbonates containing minor basinal sandstones. These contrast with cluster 375, which is made up of submarine-fan/ canyon-fill sandstones and was also placed in play 6. Note on the cluster list that cluster 373, which consists of fields in the same depositional system as cluster 375, is part of play 5 rather than play 6. Clusters 373 and 375 may also be analyzed as a combination of clusters within the same depositional system (as is done in the next section), but these clusters were divided between different USGS plays.

In this analysis, the number of fields in each grouping is almost the same: n=58 for the combination of clusters 369 and 371 and n=63 for cluster 375. For the combination of clusters 369 and 371, compared with cluster 375, the field-size/discovery- history histogram looks very different through the first and second thirds (labeled as first date, second date, etc.; see Figures C.4 and C.5). The cumulative plots for the third and fourth dates show the development of a much more filled-out pattern having the expected skew toward smaller field sizes for cluster 375, but the combination of clusters 369 and 371 shows somewhat low field numbers in classes 6 and 7, compared with cluster 375. This may mean that more fields in these size classes might be expected in the future. When the data are combined across all three classes, the cumulative plot takes on a shape in which the combination of clusters 369 and 371 appears as a "base" and cluster 375 appears to shape the frequency peaks (see Figure C.6).

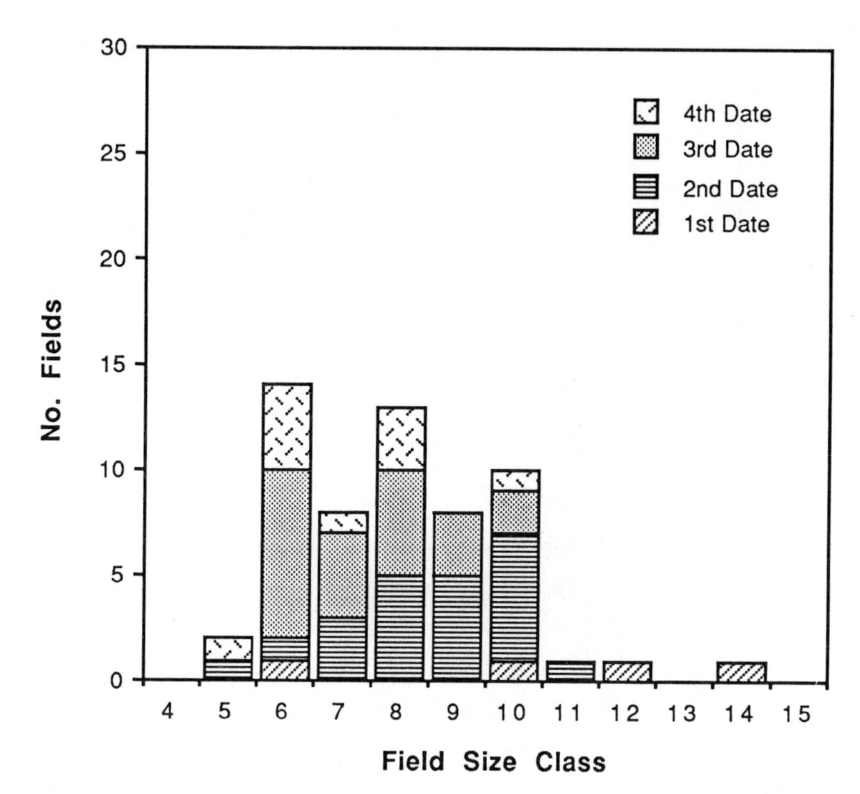

FIGURE C.4 Field-size distribution by discovery date for clusters 369 and 371. These clusters represent primarily carbonate lithologies. The histogram shows somewhat low field numbers in classes 6 and 7, compared with the histogram for cluster 375 (see Figure C.5).

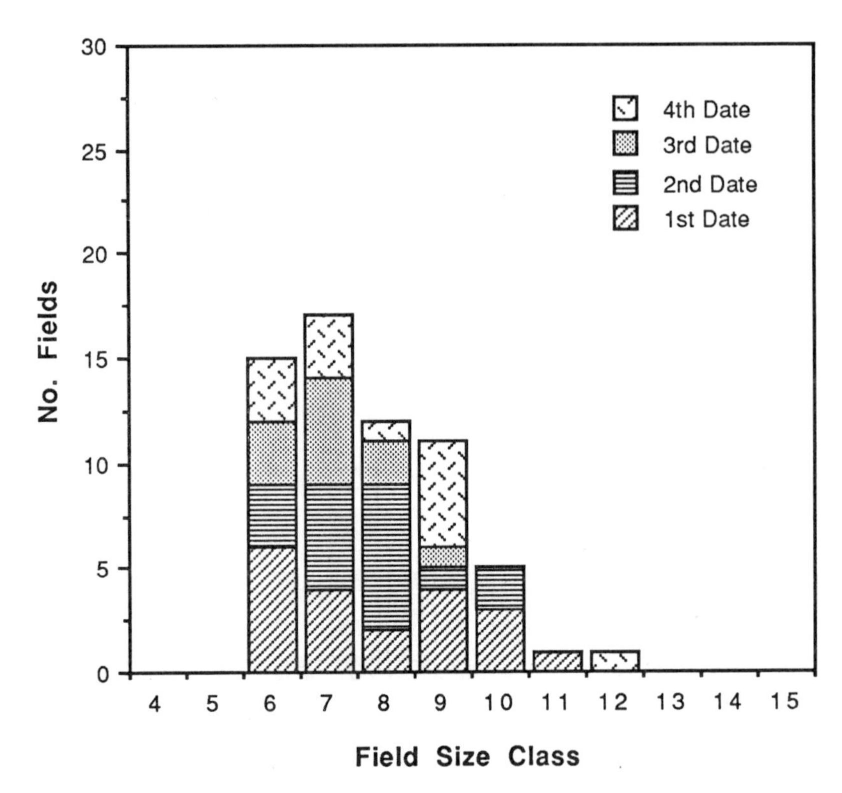

FIGURE C.5 Field-size distribution by discovery date for cluster 375. This cluster represents sandstone lithologies. The histogram is skewed toward smaller field sizes.

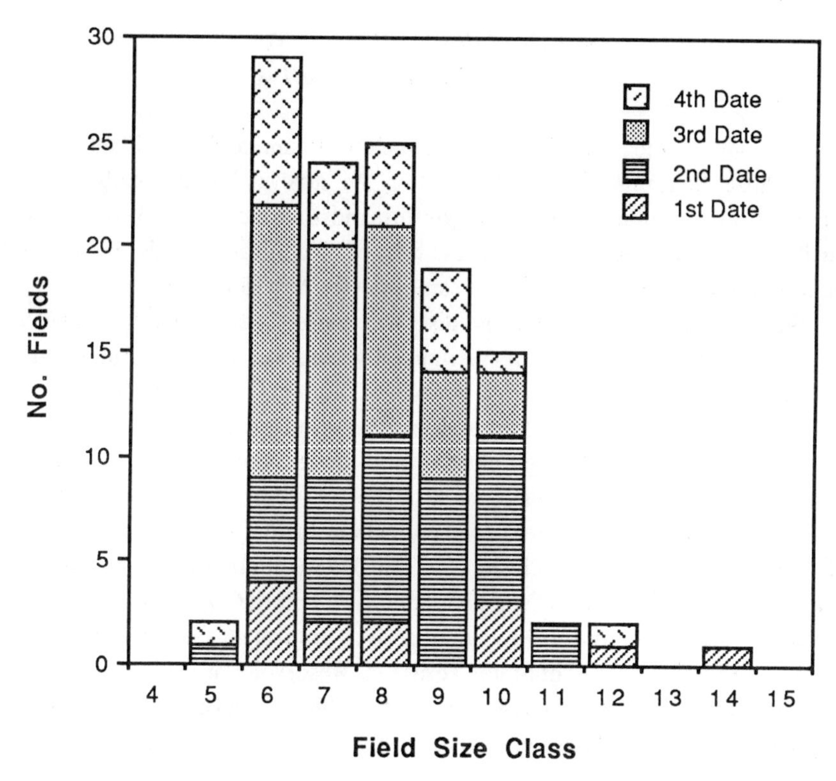

FIGURE C.6 The combined histogram for clusters 369, 371, and 375. The combination of clusters 369 and 371 appears as a base, and cluster 375 appears to shape the frequency peaks.

Analysis of Clusters 373 and 375 and the Combination of the Two

To compare the preceding plays that include dissimilar depositional systems with those that may fit together better, we analyzed individually, and in combination, two submarine-fan/canyon-fill sandstone clusters: 373 and 375. We partitioned the discovery history on a barrels of oil equivalent basis, and derived the four dates for the analysis in the same way the USGS derived them.

Patterns shown for each single discovery period are not highly similar for 373 (n=43) and 375 (n=63), but the cumulative analyses show generally similar patterns when the first and second periods of discovery are added (see Figures C.7 and C.8). Through the third and fourth periods, the histograms become quite similar for the individual clusters and for the combination of the two. Comparing the cumulative histograms of the two combined clusters with the cumulative histograms of either of the individual clusters shows more similarity, whereas the analyses placing clusters 373 and 375 in different plays show greater differences in the character of the histogram (see Figure C.9). The approach of grouping clusters 373 and 375 together derives from their similar depositional systems. Indeed, the "Atlas of Major Texas Gas Reservoirs" includes a play made up of "Upper Pennsylvanian and Lower Permian Slope and Basinal Sandstones," very similar to the combination of clusters 373 and 375 (Kosters *et al.*, 1989).

The series of histograms evaluated here shows that a play's discovery history appears different when the play is composed of clusters with different depositional systems and lithologies than it does when the clusters are more homogeneous. This is not a statistically rigorous analysis, but it demonstrates qualitatively that the combination of dissimilar depositional systems within plays could mask additional potential discoveries by suggesting a more complete distribution of discovered fields, especially toward field-size classes 6 and 7. Several large plays developed for Gulf Coast analysis may be equally subject to this phenomenon because of the heterogeneity of the depositional systems included. In the Gulf Coast, the lithology screening (carbonate versus sandstone) may not be as easily applied because these reservoirs are dominantly sandstones. In East Texas, however, the same concerns exist as raised here about the Permian Basin. The fact that the impact of grouping clusters into plays in different ways was not evaluated as part of the assessment suggests that the play content will need analysis and refinement before the next assessment.

REFERENCE

Kosters, E.C., D.G. Bebout, S.J. Seni, C.M. Garrett, Jr., L.F. Brown, Jr., H.S. Hamlin, S.P. Dutton, S.C. Ruppel, R.J. Finley, and N. Tyler. Atlas of Major Texas Gas Reservoirs. Austin, Texas: The University of Texas Bureau of Economic Geology.

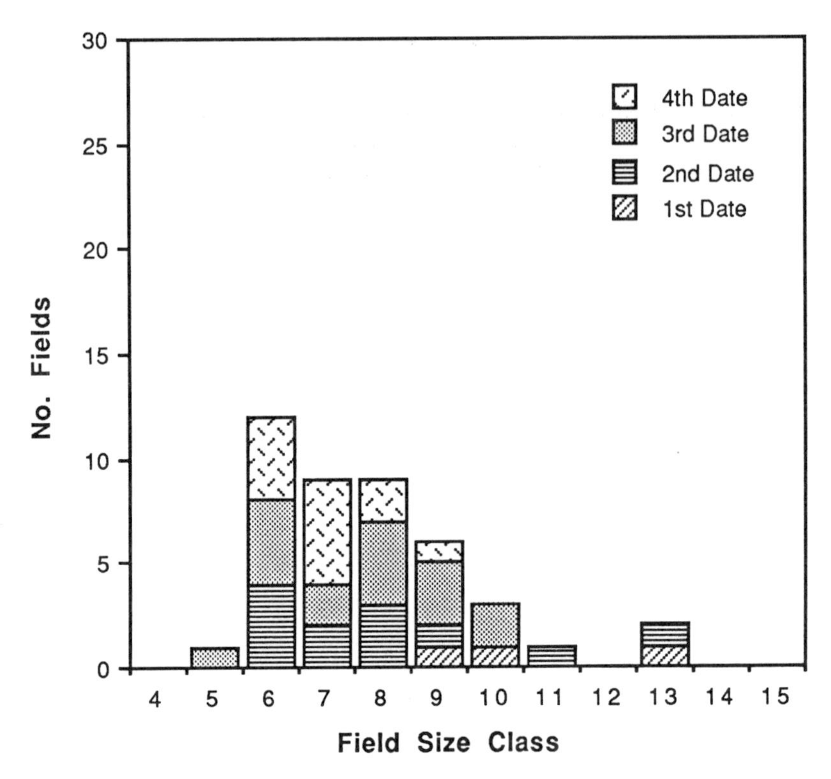

FIGURE C.7 Field-size distribution by discovery date for cluster 373, which represents submarine-fan/canyon-fill sandstone. Note the similarity in shape of this histogram to the histogram for cluster 375 (Figure C.8).

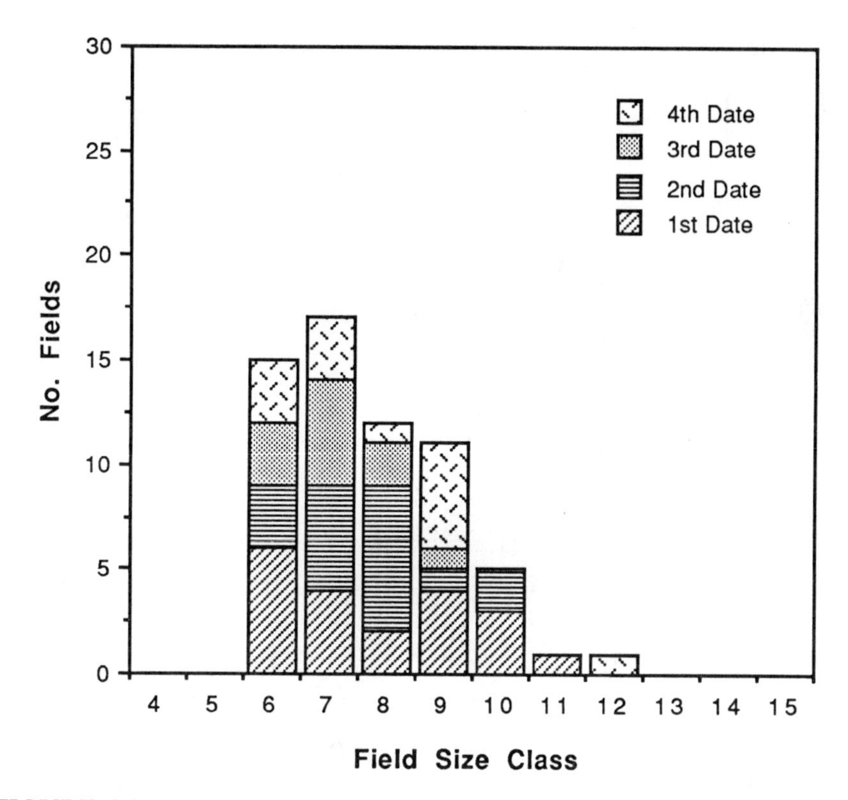

FIGURE C.8 Field-size distribution by discovery date for cluster 375, which, like cluster 373, represents submarine-fan/canyon-fill sandstone.

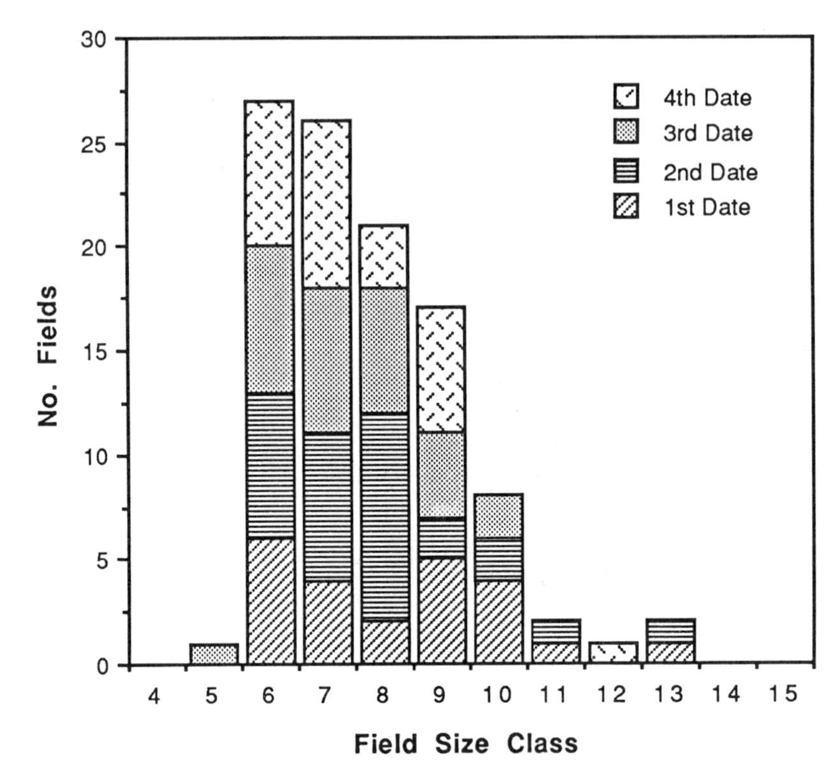

FIGURE C.9 Combined histogram for clusters 373 and 375. Note the similarity in shape of this combined histogram with the individual histograms for the two clusters (Figures C.7 and C.8).

Appendix D
Background: Components of the Petroleum Resource Base

The undiscovered conventional crude oil and natural gas evaluated in the Department of the Interior (DOI) resource assessment comprise a limited part of the total base of petroleum available for future production. This appendix defines and discusses all of the key components of the petroleum resource base for those who are unfamiliar with petroleum industry terminology.

IN-PLACE RESOURCES

The total *in-place resource base* of crude oil and natural gas (the amount that existed prior to any production) consists of the total volume that was formed and trapped within the earth's crust. The in-place resource is a function of the organic content of the source beds from which the hydrocarbons have been derived, the physical conditions under which they formed and migrated, and the effectiveness of the "trap." (A "trap" is a discontinuity in the properties of the underground formations that, because of reduction in permeability, slows the upward migration of the generated oil and gas so that they accumulate temporarily, although for a long time in human terms, and can be located and produced.)

In general, the historically evaluated recoverable portion of the in-place oil and gas resource base is composed of four main parts: *cumulative production*, *proved reserves*, *indicated* and *inferred reserves*, and *undiscovered resources* (see Figure 2.1), which will be discussed below.

Cumulative Production

The quantity of crude oil, natural gas, and/or natural gas liquids produced by a well, a field, a province, a country, or the petroleum industry from initial production to the present time is termed *cumulative production*. In the United States, crude oil is measured in terms of stock-tank barrels of 42 U.S. gallons at atmospheric pressure (14.73 pounds per square inch), corrected to 60 degrees Fahrenheit. Natural gas volumes commonly are measured in terms of cubic feet at an absolute pressure of 14.73 pounds per square inch and a temperature of 60°F.

In addition to company records, states and the federal government keep records of oil and gas production. Such records are needed for regulatory, royalty, and taxation purposes. Quantities of crude oil produced through time are reasonably well known because records have been kept since the early days of production. The records for natural gas and natural gas liquids are less complete because, through the mid-twentieth century, the market value of natural gas was small. Much of the early natural gas production was flared and no records were maintained of that production. Additionally, some states have not taxed the production of natural gas and hence have not maintained records of that production. However, reasonable estimates of cumulative production for both crude oil and natural gas have been made.

Proved Reserves

Proved reserves are those portions of in-place crude oil, condensate, natural gas, natural gas liquids, and associated substances that have been identified and are considered, on the basis of geologic and engineering data, to be recoverable under current economic and government regulatory conditions using existing technology. The producibility of those reserves is supported either by actual production or conclusive reservoir tests. Quantities of proved reserves are normally estimated using volumetric or material balance calculations or by extrapolation of production rate curves or pressure-decline curves through time.

The volume of a reservoir that is considered to contain proved reserves includes that portion delineated by drilling and defined by gas-oil, oil-water, or gas-water contacts. The adjoining portions of the reservoir that are not yet drilled, but that can be reasonably judged as economically productive on the basis of geological and engineering data, also may be considered to contain proved reserves. In the absence of information on fluid contacts, the lowest known structural occurrence of petroleum is considered to be the lower proved limit of the reservoir. The term *measured reserves*, as used by the Department of the Interior (DOI), includes that part of the identified economic resource that is estimated from geologic evidence supported directly by engineering data to be recoverable in future years from known reservoirs under existing economic and operating conditions. These measured reserves are generally equivalent to proved reserves.

Proved reserve estimates are compiled by the Energy Information Administration (EIA) of the U. S. Department of Energy (DOE). The reserve estimates are revised annually based upon a stratified sample of operator production and reserve data, additional geologic and/or engineering data, and changes in economic or regulatory conditions.

Indicated and Inferred Reserves

Indicated reserves are quantities of crude oil or natural gas that may become economically recoverable from producing reservoirs through the application of currently available improved recovery techniques. The improved techniques may already be in use, but their effects on ultimate recovery are not yet known. Alternatively, when known, the results of the application of such techniques to similar reservoirs may be used to estimate additional indicated recovery.

Inferred reserves are part of the identified economic resources that are expected to be added to proved reserves as new field wells are drilled to extend known fields, as earlier reserve estimates are revised, and as production is developed from new producing zones in known fields. The Potential Gas Committee (PGC) category of *probable resources* is similar to inferred reserves. Inferred reserves have traditionally been estimated based on statistical extrapolation of past revisions to estimates of proved reserves (growth of known fields). Estimates of probable resources are usually based on analysis of individual fields and the potential for extensions and new pool discoveries in those fields.

As inferred and indicated reserves are converted to proved reserves over time, fields increase in size. These increases collectively are termed *reserve growth*.

Inferred reserves (and indicated reserves) are not included in proved reserves due to their uncertain physical and economic recoverability and the conservatism commonly associated with initial proved reserve estimation. Neither are they included in undiscovered resources, as such quantities have in effect been identified. However, estimates of undiscovered resources must take inferred reserves into account because they are part of the ultimate recovery of a field (or pool) and as such relate to any estimate of field-size distribution in a play. Also, inferred reserves represent a significant portion of the nation's total resources. To leave them out of a discussion of oil and gas resources leaves the estimates of resources incomplete and creates a gap in the evaluation between proved reserves and undiscovered resources. Extended reserve growth must be considered in the same manner.

Indicated reserves, like proved reserves, are reported to the Energy Information Administration by some of the crude oil and natural gas operators. Inferred reserves are not reported by these operators. Rather, their existence is inferred from the historical experience that estimates of the sizes of known fields tend to increase over time. Fields grow as a result of continued drilling and the application of improved recovery techniques. The growth factor for fields of a given age is the ratio of the ultimate amount of oil or gas that the field will produce to the sum at that time of cumulative production and proved reserves. The expected future growth of fields includes inferred reserves.

An additional category of resources allied to inferred reserves is the quantities of crude oil and natural gas that might be recovered through better reservoir development. Such resources are referred to as *extended reserve growth* and are the subject of considerable current investigation. They include additional reserves to be developed through better placement of wells, known as infill drilling, and better production technology. They are based on more extensive understanding of the geology of oil and gas reservoirs, especially the discontinuities in permeability. Research on the magnitude of this type of resource is still in the early stages and there are no consistently applied methods of calculation. Many current resource estimates do not yet recognize extended reserve growth and some analysts suggest that these resources are part of inferred reserves. However, past methods of refining conventional reserve growth do not incorporate extended reserve growth, as past growth resulted from a different sort of field development. Methods are needed to estimate extended reserve growth as new technologies, and more strategic deployment of current technologies, are applied to reservoir development.

Undiscovered Resources

For long-term energy planning, it is essential to look beyond proved and inferred reserves and to anticipate the additional resources that could be developed through continued exploration for new field discoveries. This category of resources, termed *undiscovered resources*, is the principal subject of this report. Undiscovered resources are resources estimated to exist, on the basis of geologic knowledge and theory, outside of known fields and known accumulations (i.e. beyond reserves and reserve growth). Also included are resources in undiscovered pools that happen to occur within the geographic boundaries of known fields as unrelated accumulations controlled by separate structural or stratigraphic conditions (i.e., the pools are not really part of the fields).

THE DIVISION OF RESOURCES INTO CONVENTIONAL AND UNCONVENTIONAL CATEGORIES

In addition to the division of in-place resources into the four categories described above, the petroleum resource base may be divided into conventional and unconventional categories. The DOI assessment covered conventional, but not unconventional, resources. As discussed below, there are problems with limiting resource assessments to these two categories.

Conventional Resources

Resources in the *conventional* category include crude oil, natural gas, and natural gas liquids that exist as discrete accumulations in conventional reservoirs, in a fluid state amenable to extraction techniques employing the most current development practices. In the DOI assessment, conventionally recoverable resources are crude oil and natural gas accumulations that can be extracted from a well by the natural pressure within the reservoir, mechanical pumping to surface, or injection of water or gas to maintain reservoir pressure. Some commonly used techniques, such as massive hydraulic fracturing, were not considered "conventional" in the DOI's accounting. Not included were extremely viscous oil deposits, tar sands, oil shales, or natural gas either in tight sandstones or fractured shales having in-situ permeabilities of less than 0.1 millidarcy (md). Coal-bed methane, gas from geopressured brines, and gas hydrates also were not considered conventional resources. However, it is difficult to be precise about the limits of conventional resources. Some natural gas in low-permeability reservoirs is being produced and is included in some resource estimates.

The PGC divides its undiscovered category into *possible* and *speculative*. These are projected discoveries of new fields that are associated with productive formations. These new fields would be located within a productive province, and are commonly associated with productive plays but are not part of producing fields. The speculative resource category of the PGC includes projected discoveries of new fields in non-producing strata located in either producing or non-producing provinces.

Unconventional Resources

The term *unconventional resources* originally was applied to in-place resources of crude oil and natural gas not recoverable using existing or evolving technology, but which might be recoverable through the development of new technology. The boundary between conventional and unconventional resources is not well defined, because in most instances there is a continuum of geologic conditions between conventional and unconventional resources. It must be recognized that unconventional resources are being currently produced with existing technology, and, for natural gas, are making a significant contribution to current production. As improvements in recovery technology move portions of unconventional resources into the conventional category, their continued exclusion from resource assessments results in understating the volume of potentially available petroleum. Because of the evolutionary shifts of unconventional into conventional resources, it is generally difficult to distinguish between these two categories unless the technology to produce the resource does not exist and the evolutionary development of current technology does not promise a solution. The issue is complicated further through the creation of arbitrary boundaries that have been established from time to time by legislation.

There are several categories of unconventional crude oil deposits. Heavy crude oils have enough mobility that, given time, they can be produced through a well bore in response to thermal recovery methods. Tar sands contain sufficient immobile bitumen so that they will not flow into a well bore, even if thermally stimulated. Oil shale is an organic-rich rock that yields oil when heated. The oil derived from processing tar sands and shales is termed synthetic crude. It differs from conventional crude oil in that it is deficient in the lighter hydrocarbon components and commonly has a greater proportion of hydrogen-deficient (unsaturated) hydrocarbons.

A number of diverse types of unconventional gas deposits are known. Gas in low-permeability reservoirs ("tight gas") is natural gas present in blanket or lenticular sandstones that are relatively impermeable (in-situ effective permeability of less than 0.1 md). Such natural gas is widespread, especially in Cretaceous- and Cenozoic-age formations in the West (see Table D-1 for an explanation of the geologic time scale). It is now commercially recovered in some areas from the better-quality low-permeability reservoirs, mostly through the application of massive hydraulic fracturing.

The process by which vegetation is converted to coal over geologic time also generates large quantities of natural gas. Such natural gas is considered to be unconventional natural gas in the coal beds and conventional natural gas in adjacent permeable sedimentary strata. Coal-seam gas (commonly referred to as coal-bed methane) is produced in the San Juan and Piceance basins of the West, in parts of the Appalachian basin of the East, and the Black Warrior basin of the Southeast.

Shale gas was generated from organic-rich mud deposited mostly during Devonian time in a shallow sea that covered the eastern half of the United States. The gas-bearing shales are presently confined to the Illinois, Michigan, and Appalachian basins, with production further restricted, mostly to the southwestern part of the Appalachian basin.

Geopressured brine reservoirs are brine-filled rock units that have internal fluid pressures in excess of that expected from the pressure exerted by a column of water whose height is equal to the depth of burial of the reservoir. Such reservoirs commonly exist in deep, geologically young sedimentary basins in which the formation fluids (mostly brines) bear a part of the overburden load. The geopressured fluids of the Gulf Coast basin have been found to contain between 30 and 80 cubic feet of dissolved methane per barrel of brine. There is no current commercial production of this resource because the value of the extracted gas is less than the cost of producing the brine, separating the gas, and disposing of the brine.

Natural gas hydrates are ice-like crystals of water and gas (methane entrapped in voids between water crystals) which are found in regions of high pressures and low temperatures. One solid cubic foot of gas hydrates is estimated to contain about 132 cubic feet of gas. Conditions favorable to the formation of natural gas hydrates occur in permafrost zones and in relatively deep seabeds. Possibly large volumes of gas exist in the hydrate state, but there

TABLE D-1 Geologic Time Scale

Era / Period / Epoch	Major Biological Event	Began (millions of years ago)	Duration (millions of years)
Cenozoic			
Quaternary			
Holocene	Man abundant	0.01	0.01
Pleistocene	Man appears	1.6	1.59
Tertiary			
Pliocene		5.3	3.7
Miocene	Mammals diversify, grasses spread	23.7	18.4
Oligocene		36.6	12.9
Eocene	Mammals develop rapidly	57.8	21.2
Paleocene		66.4	8.6
Mesozoic			
Cretaceous	Dinosaurs become extinct, flowering plants appear	144	77.6
Jurassic	Birds appear	208	64
Triassic	Primitive mammals appear, dinosaurs appear	245	37
Paleozoic			
Permian	Reptiles appear	286	41
Pennsylvanian	Insects abundant	320	34
Mississippian		360	40
Devonian	Fish abundant	408	48
Silurian	Amphibians appear, land plants and animals appear	438	30
Ordovician	Fish appear	505	67
Cambrian	Marine invertebrates abundant	570	65
Precambrian	Simple marine plants	(±)3800	(±)3230

SOURCE: Derived from the Geological Society of America Time Scale, 1983.

is no commercial gas production from this unconventional resource in North America.

Gas in low-permeability sandstones and fractured-shale reservoirs (having in-situ permeabilities to gas of less than 0.1 md), coal-bed methane, gas in geopressured shales and brines, and gas hydrates are considered by the DOI as unconventional gas resources and were not included in the 1989 resource assessment.

RESOURCES IN RELATION TO PRODUCTION

Cumulative production, as stated previously, is the total past production, the history of which may indicate future trends. Rising or falling production trends, in the absence of major changes in reserves, economics, or geopolitics, tend to continue, at least in the short term. Proved reserves have a more direct relation to production. Although individual oil fields vary in the rates at which they can be produced, on average only about one-tenth of the remaining recoverable crude oil can be efficiently removed from a reservoir each year. Withdrawal at faster average rates commonly will cause reservoir damage that will reduce ultimate production. This average rate of recovery is sometimes expressed as the proved reserves/production ratio (R/P ratio).

The production of natural gas commonly is divided into categories of associated and non-associated natural gas. Associated natural gas is produced from a reservoir in association with oil. Non-associated natural gas is produced from a reservoir in which there is little or no oil. Some hydrocarbon liquids may be produced along with the non-associated natural gas.

The rate of production of associated gas is controlled by the oil production rate, because oil is the higher value product. Depending on reservoir characteristics, non-associated gas can be produced at faster rates than oil. Sometimes as much as one-fifth of the remaining recoverable gas can be produced in one year (R/P = 5/1). Average proved reserve/production ratios for entire regions that are undergoing intense gas development may range between 7/1 and 10/1. Large fields, with high permeability, can be produced at near maximum rates, while smaller fields, commonly with lower permeability, must be produced at slower rates. The average R/P ratio of a region is indicative of its developmental maturity, because it will consist of a combination of low R/P ratios for older, depleting fields and higher R/P ratios for newer fields that are still undergoing development. Because the larger fields are commonly found early in the exploration cycle, they will dominate production, and with depletion, tend to decrease the overall R/P ratio of the region, commonly to less than 10/1. Conversely, any proved reserves that are shut in (not being produced) or produced below their optimum production rate tend to increase the average regional R/P ratio. A regional R/P ratio much above 12/1 commonly indicates a province in which significant new discoveries are still being made and/or one in which production is not occurring at the maximum potential rate, either for technical or economic reasons.

Inferred reserves do not have as direct a relation to production as proved reserves. However, as fields grow, inferred reserves are converted to proved reserves and, as such, support production. Similarly, as undiscovered resources are converted to proved reserves, they also support production. New technology may increase the portion of in-place resources that is recoverable. If crude oil and natural gas prices rise, the increased value of the resource may encourage the utilization of more expensive production technologies, thus increasing the percentage of the resource that is recovered. Also, increased value of the resource will allow crude oil and natural gas wells to be produced for a longer time under low recovery rates, thereby increasing the total recovery.